Kingston Maurward Library
R31034
KINGSTON MAURWARD LIBRARY
WITHDRAWN

CALCULATIONS FOR AGRICULTURE AND HORTICULTURE

Farming Book Series

CALCULATIONS FOR AGRICULTURE AND HORTICULTURE

GRAHAM BOATFIELD
and
IAN HAMILTON

FARMING PRESS

First published 1984

Reprinted (with amendments) 1990

ISBN 0 85236 145 9

© GRAHAM BOATFIELD
and
IAN HAMILTON, 1984, 1990

All rights reserved. No part of this publication may be reproduced, stored in a retrieval system, or transmitted, in any form or by any means, electronic, mechanical, photocopying, recording or otherwise, without the prior permission of Farming Press Limited.

Farming Press Books
4 Friars Courtyard, 30–32 Princes Street
Ipswich IP1 1RJ, United Kingdom

Distributed in North America by
Diamond Farm Enterprises
Box 537, Alexandria Bay, NY 13607, USA

Typeset by Keyset Composition, Colchester
Reproduced, printed and bound in Great Britain by
BPCC Hazell Books
Aylesbury, Bucks, England
Member of BPCC Ltd.

Contents

Foreword

This book is one of the Farming Books series. It can be used on its own but will also be useful in conjunction with the other books in the series. It provides the basic drill needed to solve the fairly simple problems which are so often met in practice, on the farm or the horticultural unit, or in the workshop. The book also covers the calculations needed for courses in General Agriculture and General Horticulture, and for the Youth Training Schemes. It will also be helpful in many school courses.

The aim in this book is to explain methods in very simple terms, and to help the reader work through practical examples. The next step is to work out the problems in the book, for which the answers will be found at the end. After this, you can select current farm problems, using do-it-yourself measurements, current price lists, catalogues, farming papers and magazines.

This book is a help-yourself manual rather than a ready reckoner or a full-scale reference book. Further information needed, the facts and the figures, can be found in the other farming books and in magazines.

Neither farming nor horticulture is an exact science, but some sections in either of them need careful calculation and accuracy; farm machinery and the use of chemicals are two obvious examples. So in some of the following calculations the old rule of 'near enough is good enough' applies, and it is quite satisfactory to get to the nearest whole unit—kilo or square metre. In others, particularly in the machinery section, the answers need to be exact. The important things are to learn the methods; to think realistically; and to know when to make a rapid calculation that gives an approximate answer, and when you must take more time and trouble, and aim for accuracy.

Everyone today can use a calculator; they are all accurate, but some are much more complex. For the work in this book, make sure you have a simple, up-to-date calculator, in good condition. It is sensible also, from time to time, to check by other methods—to make sure you are on the right track. It is very easy to misplace a decimal point, or a zero—thus giving an answer ten times more or less than the right answer.

Simple conversions of units of measurement are sometimes needed. Often it is helpful to know a quick method which can give you the information you need; if necessary you can check it accurately later. A section of the book deals with these conversions.

Deliberately, this book refers to both Metric (S.I.) and British (Imperial) measurements. Both are in use and are likely to be used together for some time to come. Some scientists and officials would prefer an instant change to all Metric units, but the practical common man and woman will proceed at their own pace; many people are able to use both sets of measurements, to make conversions, and if necessary to mix them. It is what people have commonly had to do when travelling in another country.

The authors suggest that common sense will prevail, and that in time there will be a practical compromise. The tonne is already in general use. The metre (a long pace) is really a metric yard. The pound will become another name for half a kilo, as in France and Germany. The size of the packet or the bottle you buy in the supermarket will be determined by your convenience rather than by exact measurements. Five litres may become a metric gallon; most vehicle fuel is already sold by value rather than by measurement. It is doubtful if the hectare will become commonly used in Britain, or the quintal.

Time will provide the answer. In the meantime, be alert and make sure that your calculator is in working order.

Acknowledgements

This book is the result of the ideas and the co-operation of the two authors, who have worked together in the past in the field of agricultural education. Other ideas and examples have come from specialist colleagues who have contributed technical material. Particular thanks are due to Brian Bell NDAgrE for the machinery section; to John Pearson NDH and Ray Broughton OND(Hort) for the section on horticulture; and to Laurence Boatfield MA who has been concerned with the detailed calculations and who has made many valuable suggestions.

Chapter 1

Introduction to Calculations

Chapter 1

Introduction to Calculations

BASIC PRINCIPLES

BEFORE YOU consider the range of different calculations which might be applied to the farm or the horticultural unit, you must be sure that you understand a few of the basic principles of arithmetic, whether or not a calculator is used.

The most common operations are addition (+), subtraction (−), multiplication (×) and division (÷ or $\frac{a}{b}$).

Addition

This is appropriate where the total of a number of units or quantities is required and can be shown simply as 15 + 7 + 9 = 31.

As the numbers to be added together increase or become more complicated, it is better to list them down the page correctly one under another since mistakes are less likely to be made this way:

$$\begin{array}{r} 15 \\ 7 \\ +\ 9 \\ \hline 31 \\ \hline \end{array}$$

Subtraction

When two numbers are given and you must find the difference between them the principle of subtraction is applied. In this case the calculation normally involves only two numbers to start with:

$$124 - 39 = 85$$

or

$$\begin{array}{r} 124 \\ -39 \\ \hline 85 \\ \hline \end{array}$$

For subtraction, there is always an easy check to find out if you have obtained the correct answer. Add the bottom two numbers together in the second example and if the answer comes to the top number then it is correct.

Multiplication

This is best considered as a quicker way of carrying out additions. Let us suppose that we wanted to find out the total after adding 9 together six times. It could of course be worked out like this:

$$9+9+9+9+9+9 = 54.$$

A far simpler way would be to take the number 9 and multiply it by 6:

$$9 \times 6 = 54.$$

Just think how much quicker this is on the calculator. For the first example, twelve keys have to be pressed including the signs + and =. In the second, only four keys are pressed including × and =.

When more complicated numbers are involved it becomes even more important to understand the advantages of multiplication. It would be extremely difficult to apply the principle of addition to 9·5 × 6·2.

Division

You would use division in circumstances where a total number must be split into equal amounts. Take for example the number 12 to be split up into four equal amounts. It would be written as:

$$12 \div 4 \quad or \quad \frac{12}{4}.$$

In either case the answer is 3. Once again more complicated numbers give rise to more difficult calculations but the principle remains the same.

Combining the Arithmetic

Quite frequently a calculation is required which will use more than one of the four operations already described.

Take for example $7 \times 6 + 2$.

It is not apparent which of the two operations should be carried out first. It must be worked out correctly otherwise an entirely misleading answer could be obtained.

For the above, the choice is:

either $7 \times 6 = 42, \quad 42 + 2 = 44$

or $6 + 2 = 8, \quad 8 \times 7 = 56.$

Sometimes guidance is given by putting brackets around a set of numbers and this means 'work out the answer inside the brackets first'.

Therefore $(7 \times 6) + 2 = 44$

but $7 \times (6 + 2) = 56$

and here both answers are correct.

Where no guidance is given as to the correct order of calculation, a helpful memory jogger which has stood the test of time is:

B	O	D	M	A	S
Brackets	of	÷	×	+	−

to be calculated in that order as and when they might appear. Applying this rule to the example given at the top of this section, the correct answer can only be:

$$7 \times 6 + 2 = 42 + 2$$
$$= 44.$$

Some of the more advanced and expensive calculators will sort out the calculations in the correct order. If the calculator being used does not sort the order, it will be important to follow the rule given above.

Decimals

Not all numbers used in calculations are whole numbers. More often than not they are part of a whole number or a combination of a whole number and a part of one.

The part of the whole number can be expressed either as a fraction or a decimal:

$$12\tfrac{1}{4} \quad \textit{or} \quad 12{\cdot}25.$$

However, modern calculators are designed to deal with numbers expressed in decimal form and it is intended throughout this book to ignore fractions other than to show how they can be converted to decimals.

Fractions are always quoted as two numbers, one above the other separated by a line. For example, a half is $\frac{1}{2}$, a quarter is $\frac{1}{4}$, and an eighth is $\frac{1}{8}$.

In these examples and in every other case involving a fraction, divide the bottom number into the top number and the decimal equivalent is obtained.

Therefore: $\frac{1}{2}$ is $1 \div 2 = 0{\cdot}5$

$\frac{1}{4}$ is $1 \div 4 = 0{\cdot}25$

$\frac{1}{8}$ is $1 \div 8 = 0{\cdot}125$

$\frac{2}{5}$ is $2 \div 5 = 0{\cdot}4$.

Earlier, $12\frac{1}{4}$ was referred to. This now becomes $12 + 0{\cdot}25 = 12{\cdot}25$.

Decimal numbers after the decimal point can go on and on. Such precision is not often required in agriculture or horticulture. A convenient and acceptable way of shortening or abbreviating the number is to correct it to one, two or three places of decimals.

When correcting to a stated number of decimal places, look at the next figure beyond and if it is 5 or above add one more to the figure before and forget the rest.

Example: Take the number 7·1425926.

Corrected to one place of decimals, it becomes 7·1 because 4 is less than 5.

Corrected to two places of decimals, it becomes 7·14 because 2 is less than 5.

Corrected to three places of decimals, it becomes 7·143 because the fourth figure after the decimal point is 5.

Corrected to four places of decimals, it becomes 7·1426 because 9 is greater than 5.

When working out problems in decimals, it is most important for the decimal point to be inserted in the correct place. For addition and subtraction, always keep the decimal points in line with each other and make sure the decimal point in the answer appears in line with the others.

Examples:

63·5 + 0·25 + 8·75 + 118·125

becomes

```
   63·5
    0·25
    8·75
+ 118·125
---------
  190·625
---------
```

11·7 − 8·07

becomes

```
  11·7
− 8·07
------
  3·63
------
```

Multiplication and division are much more difficult to work out on paper and using a calculator can save a great deal of time. The exception to this is when multiplying or dividing by 10, 100, 1,000 etc., because all that is required is to move the decimal point to the right or left.

Question: How many places should the decimal point be moved?
Answer: By as many places as there are 0s in the multiplier or divider, i.e. 10 has one 0, 100 has two 0s etc.
Question: In which direction should the decimal point be moved?
Answer: To the right for multiplication; to the left for division, making the number larger or smaller respectively.

Example: take 128·321 as the number to be multiplied or divided:

128·321	*×10*	*×100*	*×1,000*
	1,283·21	12,832·1	128,321
128·321	*÷10*	*÷100*	*÷1,000*
	12·8321	1·28321	0·128321.

Incidentally with decimal numbers lower than 1, it is a good habit to put a zero in front of the decimal point since the point can sometimes be overlooked or is so indistinct as to give a totally false value to the number.

Averages

There are many examples in farming and horticulture where averages are calculated. In a given set of conditions a number of results might be obtained, some of which are high and others lower. To obtain some idea of the average performance, all the results are added together and divided by the number of results used.

Example: Ten cows gave the following daily yields in kg of milk in the week following turnout to grass in the spring:

10; 15; 8; 20; 18; 17; 17; 13; 15; 16.

What is the average yield?

Added together, these yields total 149 kg and from ten cows this makes an average yield per cow of $149 \div 10 = 14{\cdot}9$ kg of milk.

Percentages

A percentage is written as %. The calculation of percentages is used for many different purposes. It is simple to understand if you remember that when a percentage figure is stated, it really means so many parts in 100. Thus, 15% means fifteen parts in one hundred.

Examples: (a) $15\% = \frac{15}{100} = 0{\cdot}15$

(b) $22\frac{1}{2}\% = \frac{22{\cdot}5}{100} = 0{\cdot}225$

(c) $120\% = \frac{120}{100} = 1{\cdot}2.$

Supposing therefore you wanted to find each of the above percentages as applied to the number 60:

(a) $15\% \text{ of } 60 = 0{\cdot}15 \times 60 = 9$

(b) $22\frac{1}{2}\%$ of 60 = $0{\cdot}225 \times 60 = 13{\cdot}5$
(c) 120% of 60 = $1{\cdot}2 \times 60 = 72$.

This shows that you do not need to have a calculator with a percentage key to be able to work out percentages.

THE CALCULATOR

Choosing a Calculator

There is a considerable range of calculators available to buy these days. Some are very expensive and are capable of performing calculations beyond the range of those likely to be made on farms or horticultural units.

You should make sure that the calculator you use can perform the basic operations of +, −, ×, ÷ and =. It should have a 'floating decimal point' to ensure that the point always appears at the right place in the answer, so long as the calculator is used correctly. If you press the wrong key—which is easily done—to avoid starting the calculation all over again, there is a key which just cancels the last entry (labelled C or C/E). Another useful feature is the % key, and you should study in the instruction book how this should be used.

A calculator is useful to make conversions—such as those from Imperial to Metric—although there are also tables and book references which do the same job. (See pages 106–7.)

Using the Calculator

Each model will carry its own set of instructions but calculators are usually programmed to work out answers in the same way as the problem is written down on paper.

Examples:

15 plus 7 plus 9 equals 31 ($15 + 7 + 9 = 31$)

124 minus 39 equals 85 ($124 - 39 = 85$)

9 multiplied by 6 equals 54 ($9 \times 6 = 54$)

÷ =
12 divided by 3 equals 4

× + =
7 multiplied by 6 plus 2 equals 44

÷ =
2 divided by 5 equals 0·4

× % =
60 multiplied by 15 per cent equals 9

× =
60 multiplied by 0·15 equals 9

} in fact the % key has been programmed to divide the answer by 100

Do not forget to cancel one set of calculations before starting another. If you do not, the calculator will still keep on using the previous information—and this leads to confusion.

Finally, always remember to carry out calculations using the same units. For example, you would not use grams and kilograms, or metres and centimetres, in the same calculation; you should use either one or the other, converting if necessary. This applies to any other values, whether in metric or imperial. Use the same ones in the same sum.

Such conversions are very easy in metric, as it is based on systems of 10. You add a 0, take away a 0, or move a decimal point.

Note: Make sure that the calculator you use is a current model and up to date. Avoid old ones, which may operate on a different system. Always refer to the handbook or the sheet of notes provided with the calculator. Then you can use it to the full.

Take care that your calculator batteries are not getting weak. This can distort the results.

General Problems

Where appropriate, correct all answers to two places of decimals. You should attempt to answer questions 1 and 2 without using a calculator.

1. *Add:* 25·4 + 0·987; 6·375 + 10·0875; 51 + 0·075.

2. *Subtract:* 14·6 − 7·08; 0·01 − 0·001; 51·1 − 25·579.
3. The following is a table of dimensions for different sized rectangular blocks and uses the formulae for areas and volumes referred to in Chapter 2. The answers are obtained by multiplication and division. Complete the missing answers.

Length	*Breadth or width*	*Height*	*Area of base*	*Volume of block*
?	5·81 cm	?	59·38 cm^2	377·66 cm^3
8·5 m	4·75 m	?	?	161·52 m^3
?	5·13 mm	3·14 mm	61·56 mm^2	?
2·75 m	?	?	2·06 m^2	1·24 m^3
?	4·44 cm	6·31 cm	25·22 cm^2	?

4. Express the following fractions as decimals:

   $\frac{5}{8}$; $\frac{35}{100}$; $3\frac{4}{9}$; $70\frac{8}{27}$; $\frac{61}{1000}$.
5. If 1 inch equals 2·54 cm, express $\frac{5}{8}$ of an inch in cm.
6. Calculate the following: $\dfrac{3{\cdot}225 \times 6{\cdot}97 \times 10{,}000}{7{\cdot}55 \times 2{\cdot}224}$
7. Calculate 6·25 + 4·98 × 7·31 ÷ 0·78 − 10·2.
8. The diameter of a metal rod was measured in millimetres at different points along its length, with the following results:

   0·729; 0·698; 0·706; 0·747; 0·732; 0·681; 0·691.

   Find the average diameter.
9. The revolution counter of a tractor reads more than the actual speed of the engine by 2·25%; there is a known error in the counter. What is the real speed of the tractor engine when the counter reads 2,200 revolutions per minute?
10. (a) An increase of 10·8% has been awarded on a salary of £5,500 per year. How much will the new salary be for the next year?
    (b) If overtime rates increase from £2·25 to £2·65 per hour, is this the same percentage increase as in (a)?

Chapter 2

Farm Crop Calculations

Chapter 2

Farm Crop Calculations

AREA

IN FARMING it is often necessary to find the area of flat surfaces—a field, a roof, floor space, a wall. Area is measured in square units—square millimetres (mm^2), square centimetres (cm^2), square metres (m^2), hectares and, in British units, square yards and acres.

It is simple to calculate the areas of certain basic shapes, as follows. But make sure that all measurements are in the same units; multiply millimetres by millimetres or metres by metres.

Rectangles and Squares

Shapes with four straight sides and four right angles. In rectangles the opposite sides are equal in length; in squares all four sides are equal. In either case the total area is found by multiplying together the lengths of two adjacent sides.

Length multiplied by Breadth, or $L \times B$.

Parallelogram

A shape where opposite sides are parallel, and equal, but there are no right angles; in fact, it is like a rectangle squashed over sideways. You find the total area by multiplying together the length of the base and the height.

Length multiplied by Height, or $L \times H$.

Triangle

A three-sided shape. You find the total area by multiplying together the base and the vertical height, and dividing by 2. (It is exactly the same if you say $\frac{1}{2}$ Base × Height or $\frac{1}{2}$ Height × Base.)

$$\frac{\text{Base multiplied by Height}}{2} \text{ or } \frac{B \times H}{2}$$

Circle

The area of a circle is often needed in dealing with circular tanks or silos; you need to find the area of the base (which is a circle) before you can find out how much it will hold. In some countries, stacks are made circular, and the same calculation has to be done.

A straight line right across a circle, passing through the centre, is the Diameter. Half this distance (centre to outside) is the Radius. The distance round the outside of the circle is the Circumference. In these calculations, we use the factor π (called pi) which is roughly $\frac{22}{7}$ or 3·14.

You find the Circumference by this method:

$$\text{Diameter} \times \pi \quad \text{or } D \times 3{\cdot}14 = \text{Circumference.}$$

You find the Area by this method:

$$\text{Radius squared} \times \pi \quad \text{or } R^2 \times 3{\cdot}14 = \text{Area.}$$

Example: Diameter of the bottom of a silo is 5 metres. Find the Circumference and the Area.

5 × 3·14 = 15·7. The Circumference is 15·7 metres.

If Diameter is 5 m, Radius is 2·5 m, Radius squared is 2·5 × 2·5 = 6·25 m^2.

6·25 × 3·14 = 19·63. The Area is 19·63 square metres.

Area of Land

Formerly land was measured in square yards or acres. Even today many farmers still prefer to use acres because they judge this unit to be of a more sensible size than the hectare which is the metric alternative. However, much of the farming press, the advisory services and some industries associated with agriculture express their recommendations and results in metric terms. Here are some useful equivalents. Further conversions are given at the back of the book.

4,840 square yards = 1 acre

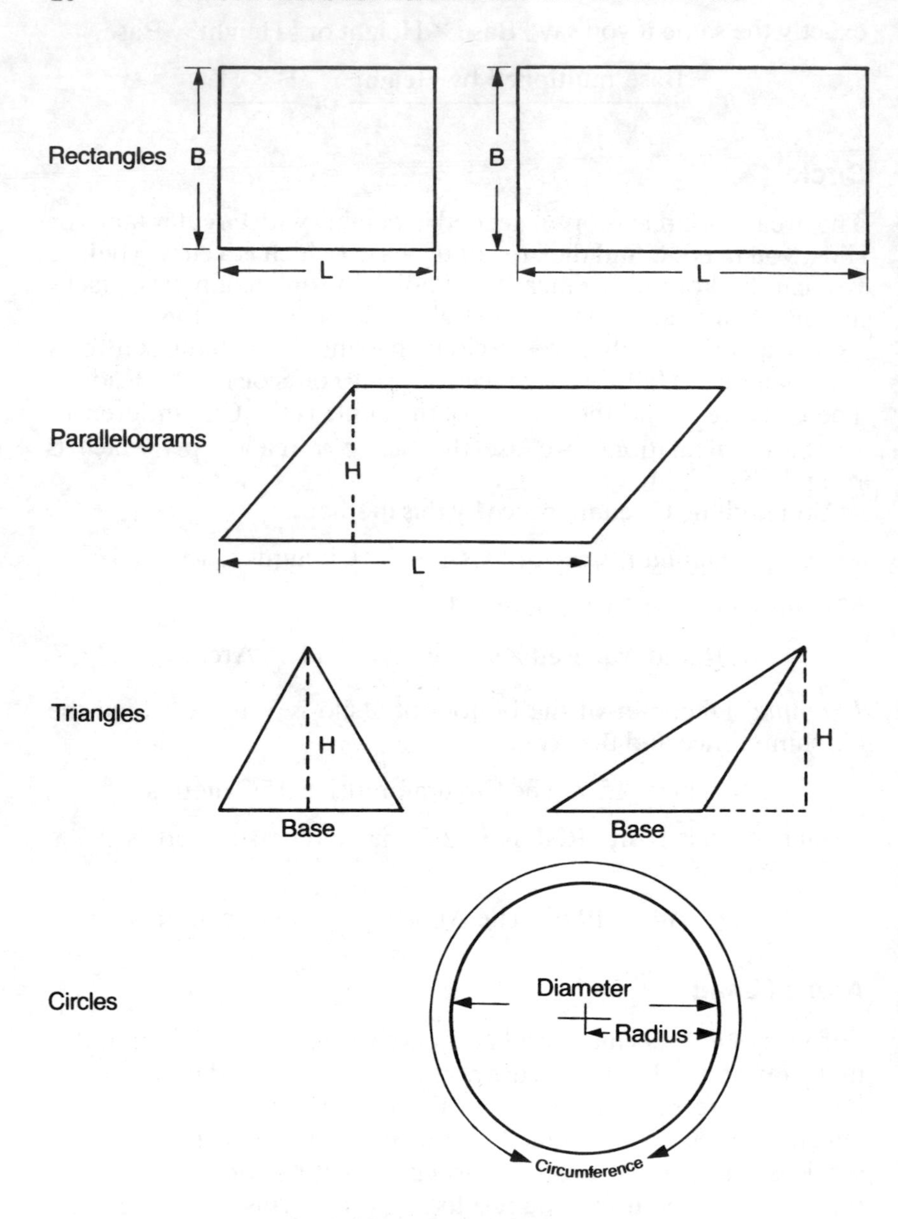

CALCULATING AREAS

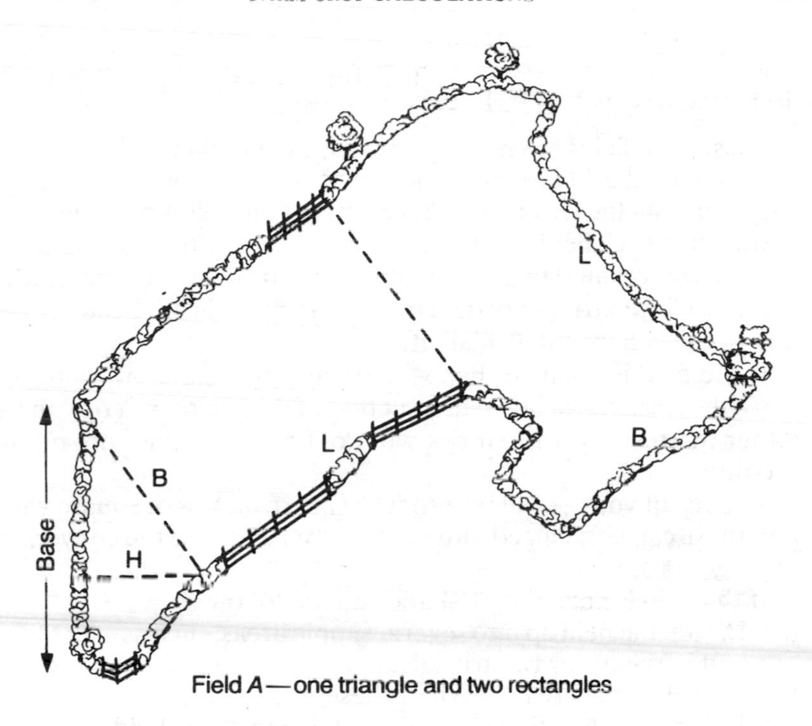

Field *A*—one triangle and two rectangles

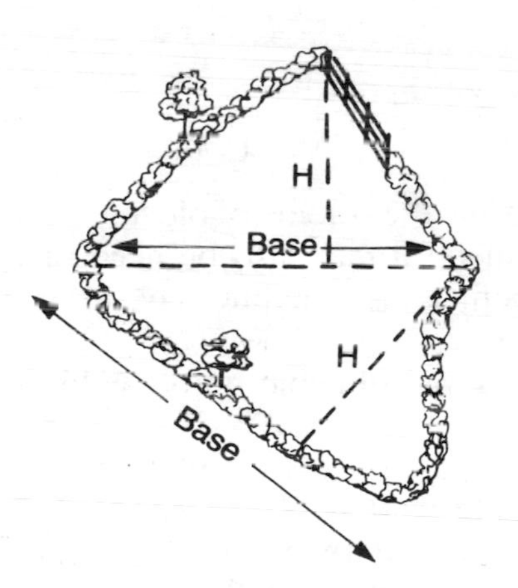

Field *B* — two triangles

CALCULATING FIELD AREAS

2·5 acres (approx) = 1 hectare
1 square yard = 0·84 square metre (m^2)
10,000 square metres = 1 hectare.

Most farm fields are not regular shapes, and their sides (hedges, fences or ditches) twist and bend about. In many cases, you can get the area from the Ordnance Survey maps (1 in 2,500 maps) where it is shown in each field by the figure below the Field Number.

When you cannot find the area properly from the Ordnance Map (because the map is out of date or the original field has been divided up) you may have to calculate it.

If the field is a simple shape—a square, rectangle, or triangle—work it out as shown in the first section of this chapter. You can use either metres or yards (strides will do if accuracy does not matter too much).

The result you get will be either in square metres or square yards and these can be changed into hectares or acres (see the conversion on page 106).

If the field is not a simple shape, do the following:

(*a*) Divide the field up into several simple areas, such as a rectangle and a triangle or two triangles.
(*b*) Calculate the area of each of these.
(*c*) Add the lot together to get the total area of the field.

VOLUME

We often need to find the volume of blocks, stacks, tanks, buildings and other containers. It may also be necessary to calculate what they contain, whether it is concrete, hay, straw, potatoes, water or diesel fuel.

Volume is measured in cubic units—cubic millimetres (mm^3), cubic metres (m^3). These can be related to measurements of capacity—litres—and also to weights of material. (See pages 25 and 29 for more information.)

It is simple to calculate the volumes of the basic shapes. Make sure that all measurements are in the same units.

Square or Rectangular Block such as a barn full of hay or straw, a stack, a building, or a piece of concrete.

Multiply Length by Breadth by Height. $L \times B \times H$.

Example: We want to lay a concrete floor which measures 9 metres by 4 metres, to a depth of 150 millimetres. We are buying-in ready-mixed concrete; how much should we order?

Convert the millimetres to metres and work it out in cubic metres, which is the normal measurement for concrete.

$9 \times 4 \times 0{\cdot}15 = 5{\cdot}4$ cubic metres ($5{\cdot}4\ m^3$).

Triangular shape such as a clamp of roots.

Multiply Length by Breadth by Height, and divide by two.

$$\frac{L \times B \times H}{2}$$

Cylinder such as a silo or circular tank.

Multiply the area of the base by its height.

$$\pi r^2 \times H$$

(If the cylinder lies on its side multiply the Area of the circular end by the cylinder's length.)

Example: What is the capacity of a cylindrical tank measuring 2·2 metres across (diameter) and 2·5 metres high?

If the diameter is 2·2 metres the radius is 1·1 metres.

$$3{\cdot}14 \times 1{\cdot}1 \times 1{\cdot}1 \times 2{\cdot}5 = 9{\cdot}5 \text{ cubic metres } (m^3)$$

$$[\pi r^2 \text{ (the Base)} \times H]$$

1 cubic metre equals 1,000 litres (see page 107), so the capacity of the tank in litres is $9{\cdot}5 \times 1{,}000 = 9{,}500$ litres.

Cone Shape such as the roof of a round stack.

In this case multiply the Area of the Base by the Height and divide by three.

$$\frac{\pi r^2 \times H}{3}$$

STORAGE OF MATERIALS

Many crops are harvested and stored on the farm in some way—either for sale later, or for feeding to livestock. Examples are cereal grains, peas and beans, potatoes, feeding roots, silage, hay, straw and also farmyard manure. It is often necessary to calculate just how much is stored, so that it can be offered for sale or used in daily rations (see the table on page 80 of *Farm Livestock*).

Grain

Much of the grain which is combine-harvested is stored for a time on the farm in bulk—either in square or circular containers, or in heaps on the floor. The figures given below show the average weight of grain in one cubic metre and the number of cubic metres which on average will be taken up by one tonne of grain. Using them it is easy to work out the actual weight of grain from the size of the heap or the measurements of the silo.

Grain	*kg/m³*	*m³/tonne*
Wheat	785	1·3
Barley	705	1·4
Oats	520	1·9
Beans	840	1·2
Peas	785	1·3
Oilseed rape	690	1·4
Ryegrass seed	317	3·2

The method for finding the storage capacity of a silo or other container is as follows:

(a) Find the area of the base, whether it is square or round (see pages 18 and 19). Take all measurements in metres.
(b) Multiply this figure by the height of the container, which gives full potential capacity, or by the height of the grain in it, which gives the actual quantity present.
(c) To obtain the weight of grain from the cubic capacity (cubic metres, m^3), either
  (i) divide by the figure shown in the second column of the table for type of grain, or
  (ii) for slightly greater accuracy, multiply by the figure shown in the first column of the table to obtain the weight in kilograms and then divide by 1,000 to convert to tonnes.

Example: A grain silo measures 4·9 metres by 4·4 metres and the grain in it is 3 metres high. How many tonnes of wheat are there?

Cubic capacity in m^3 is $4{\cdot}9 \times 4{\cdot}4 \times 3 = 64{\cdot}7\ m^3$

the weight by method (i) is:

$$\frac{64{\cdot}7}{1{\cdot}3} = 49{\cdot}8 \text{ tonnes}$$

or by method (ii):

$$64{\cdot}7 \times 785 = 50{,}789 \text{ kg} = 50{\cdot}789 \text{ tonnes.}$$

Remember that the weight of grain per cubic metre will vary according to the moisture content and the figures given in the table can be regarded only as a guide.

Grain harvested from the field can often be of a higher moisture content than is safe for long-term storage. It will need to be dried and in so doing it will lose weight. Its dried weight will, however, be what is sold and paid for. Using a very simple formula, a conversion can be calculated to obtain the final dried weight.

It is:

$$\text{tonnes of grain harvested} \times \frac{(100 - \text{moisture content as harvested})}{(100 - \text{moisture content when dried})}$$

$$= \text{weight of dried grain in tonnes.}$$

Example: 30 tonnes of barley was harvested at 19% moisture and is to be dried down to 16% before selling. What will its dried weight be?

$$30 \times \frac{(100 - 19)}{(100 - 16)} = 30 \times \frac{81}{84} = 28{\cdot}9 \text{ tonnes of barley.}$$

Liquids

For all liquids 1 cubic metre equals 1,000 litres.

Weight per litre

Water	1·00 kg	Petrol	0·75 kg
Milk	1·03 kg	Paraffin	0·80 kg
		Diesel fuel	0·84 kg

A very simple calculation is that dividing any given figure stated as pounds per gallon by ten converts it into kilogrammes per litre.

Hay and Straw

It is common now for these materials to be baled and stored in buildings and barns. In these cases the amount stored can be obtained by calculating the volume and using the figures on page 28. Hay and straw are also stored in field stacks of different shapes; the contents can be worked out in the following way.

Rectangular stacks

(a) Allow for loose ends, foundations and top (if topped up with loose straw). Take measurements in metres.
(b) Measure height from the ground to the top (if a flat top) or to a point halfway up the roof (if a gable roof). Measure length and breadth midway between ground and eaves.
(c) Multiply length × breadth × height, which gives the volume of the stack in cubic metres.
(d) Divide this figure by the number of cubic metres per tonne, as shown in the table on page 28. The final figure shows the number of tonnes in the stack.

Example: A flat-topped stack of barley straw bales measures 8 metres by 5 metres with a height of 4 metres. Multiply together $8 \times 5 \times 4 = 160\ \text{m}^3$. Divide by 11·5 for baled barley straw:

$$\frac{160}{11{\cdot}5} = 13{\cdot}9 \text{ tonnes of straw.}$$

Circular stacks

These are calculated in the same way, but the necessary measurements are the radius and the height. Multiply π by the radius squared by the height, which gives the volume in cubic metres. Divide this by the number of cubic metres per tonne according to the type of material.

Example: A circular stack of loose hay measures 5·4 metres across with a height of 6·4 metres. Multiply

$$\frac{5{\cdot}4}{2} \times \frac{5{\cdot}4}{2} \times 3{\cdot}14 \times 6{\cdot}4 = 146{\cdot}5 \text{ cubic metres.}$$

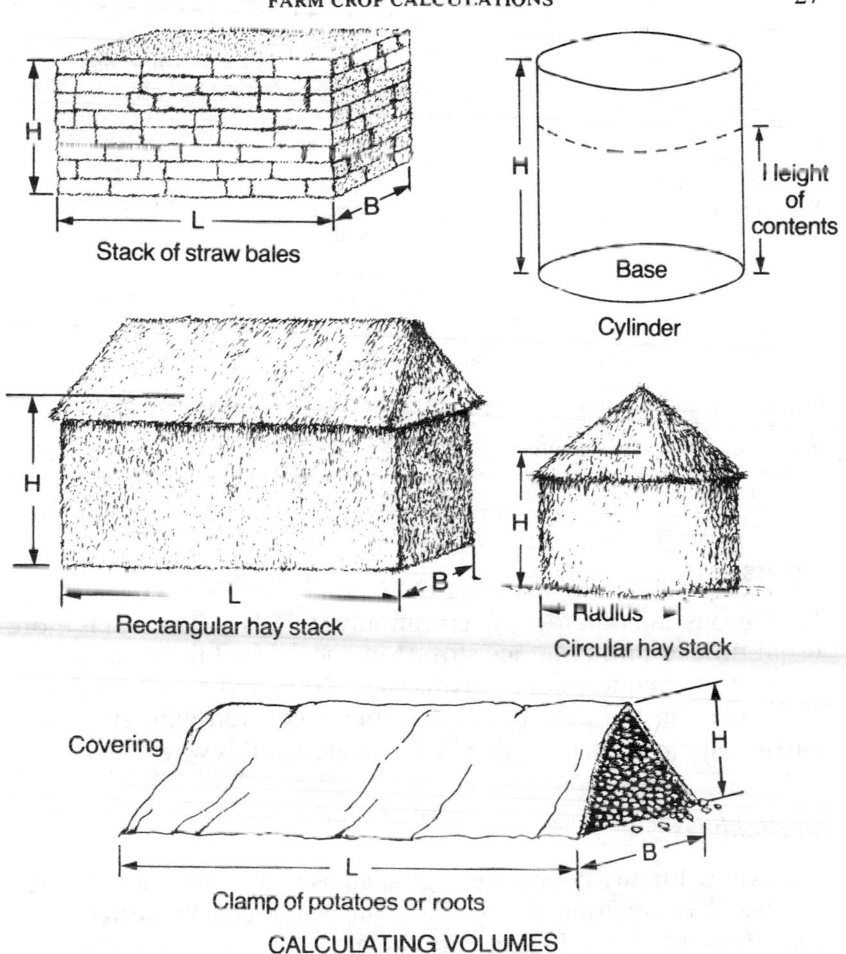

CALCULATING VOLUMES

Divide this by 9·2

$$\frac{146 \cdot 5}{9 \cdot 2} = 15 \cdot 9 \text{ tonnes of hay.}$$

Silage (and to a lesser extent grain) is not always stored at the same dry matter content. The lower the dry matter, the fewer cubic metres of it there will be to the tonne. For example:

Silage at 20% dry matter — 1·38 cubic metres per tonne
30% dry matter — 1·62 cubic metres per tonne

and intermediate figures can be estimated proportionately.

Material		*Cubic metres per tonne*
Hay	baled	6·0
	loose	9·2
Wheat straw	baled	9·7
	loose	17·6
Barley straw	baled	11·5
	loose	22·8
Oat straw	baled	10·5
	loose	19·1
Silage at 25% dry matter		1·5
Well-rotted farmyard manure		1·15

Roots

Root crops and potatoes are commonly stored on farms. It is more usual now to find potatoes stored in special buildings, and other roots are sometimes stored in this way. Roots and potatoes are also still stored in the field clamp (or 'pie'), covered with straw and earth. Quantities in store can be calculated in this way:

In Buildings

(a) Allow for any floor covering (such as straw), any lining to the walls, or covering on top. Take measurements in metres.
(b) Measure length, breadth and height.
(c) Multiply these measurements together, which gives the volume in cubic metres.
(d) Divide this figure by the number of cubic metres per tonne, as shown in the table above. This final figure is the number of tonnes in the store.

In Field Clamps (of the traditional triangular shape)

(a) Allow for earth and straw coverings and foundations.
(b) Measure length at one-third up the clamp. Measure average breadth and height.
(c) Multiply these together and divide by two, which gives the volume of the clamp in cubic metres.

(d) Divide this figure by the number of cubic metres per tonne, as shown in the table below. This final figure is the number of tonnes in the clamp.

This shows the method clearly:

$$\frac{\text{Length} \times \text{Breadth} \times \text{Height}}{2} = \text{volume of clamp in m}^3$$

$$\frac{\text{Volume in m}^3}{\text{m}^3 \text{ per tonne}} = \text{tonnes in clamp.}$$

Roots	*Cubic metres per tonne*
Potatoes	1·56
Mangels, sugar beet, fodder beet, swedes	1·75
Turnips	1·90
Carrots, parsnips	2·10

CROPS—SEEDS AND YIELDS

Buying and selling is now conducted in kilograms and metric tonnes but many farmers think in terms of pounds, hundredweights and tons and relate the performance of crops to acres. This section will concentrate on metric terms but will also give some helpful conversions.

The relationship between grams, kilograms and tonnes is that each unit is one thousand times larger than the one below.

i.e. 1 gram × 1,000 = 1 kilogram
1 kilogram × 1,000 = 1 tonne.

Conversion from one to another becomes easy if you remember the rules of multiplying and dividing by 10, 100, 1,000, etc. (See page 13.)

Farmers are now persuaded of the advantages of drilling and planting seeds to achieve the optimum number of plants in a given area. Too few plants will allow gaps which waste land and permit weeds to creep in and take over. Too many plants will mean that they are so close together that they compete with each other for available moisture, nutrients and light to the detriment of final yield.

Current recommendations for seed rates take into account the need for an optimum plant population and also draw the distinction between plump seeds obtained previously from a good harvest or thin seeds from a bad harvest.

Thousand Grain Weight (T.G.W.)

A thousand plump grains will weigh more than 1,000 thin grains. To put it another way a greater weight of plump seed will have to be sown to drill the same number of seeds to the hectare.

Example:

T.G.W. in grams	*Kg of seed per ha to obtain same plant population*
36	144
40	160
44	176
48	192

Notice that for every increment of 4 grams in T.G.W., the seed rate increases by 16 kg per hectare, so you can work out seed rates from any other T.G.W.

Seed Yields and Losses

Not all seeds sown will germinate or emerge, either because they are infertile or have not been drilled correctly (too deep or shallow, too wet or dry). Some are taken out by birds and other pests and on autumn-sown crops there will be some 'winter kill' of plants during the winter months, due to harsh weather conditions and other risks.

Most seed sold in this country is of guaranteed germination percentage, for example 95% germination, which means that up to 5% of the seed may fail to germinate. As the seedlings grow you can expect a further loss of something like 7·5% of the original sowing; this is known as 'post-drilling loss'.

For winter cereals the target is 400 seeds drilled per square metre, which allows for 350 plants per square metre to emerge.

This is calculated as follows: 5% loss at germination plus 7·5% loss post-drilling gives a total loss of 12·5%; or an emergence of 87·5%.

$$400 \text{ seeds per m}^2 \times \frac{87{\cdot}5}{100} = 350 \text{ plants per square metre.}$$

Winter kill can cause a further loss of 10–20% of the plants, depending on the severity of the winter. Allowing for a 15% loss, the plants expected to survive through into the spring would be:

$$350 \times \frac{85}{100} = 298 \text{ plants per square metre.}$$

Normally the seed rate for spring-sown crops can be reduced by up to 10% below the rate for those sown in the winter because of the reduced hazards and losses.

Crop	*Sowing rate (kg per hectare)*	*Yield (tonnes per hectare)*
Winter wheat and barley	180	6·0
Spring barley	160	5·0
Oats	200	4·8
Rye	200	3·0
Field beans	250	3·0
Peas	250	3·5
Oilseed rape	8	3·0

Simple conversions

(a) Divide the sowing rate in kilos per hectare by two, and then deduct 10% of the original figure, to give the sowing rate in *kilos per acre*.

e.g. Winter wheat:

$$\frac{180}{2} - 18 = \quad 90 - 18 = 72 \text{ kg per acre.}$$

(b) Yield in tonnes per hectare multiplied by eight will give the yield in *hundredweights per acre e.g.* Spring barley: 5·0 × 8 = 40 cwts per acre.

Checking on plant populations in the spring helps to determine the policy of top dressing during the growing period. There are normally two common drill widths: 180 mm and 110 mm. To obtain the number of plants in a square metre, count the plants in the appropriate length of a row, as shown in the table—but remember to take a number of sample readings across the field and obtain an average.

Drill width	*Length of one row for 1·0 m^2*
180 mm	5·5 metres
110 mm	9·1 metres

If the crop is fairly even then you only need to count the plants in half of the above length but you must multiply each answer by two.

Roots and Forage Crops

Crop	*Seed rate (kg per ha)*	*Yield (tonnes per ha)*
Sugar beet	13	40
Mangels and fodder beet	10	70
Swedes	4	50
Turnips	4	40
Kale	5	60
Potatoes: early	4.0 tonnes	15
maincrop	3·0–3·5 tonnes	35

For crops grown in rows the yield per hectare can be estimated by weighing a few samples taken from 1 square metre at various places across the field.

Here is a method to follow:

(a) Find out the drilling/planting width in metres or convert to metres. This will almost always be less than one metre.
(b) Divide this drill width into 1 to obtain the length of one row which should be harvested to give the weight of crop per square metre.

(c) If the crop is to be grazed, e.g. kale or stubble turnips, then only harvest that part of the crop the animal is likely to eat above the ground. For sugar beet, remove the top and retain the saleable root.
(d) Add up the sample weighings and obtain an average in kg per square metre across the field.
(e) Multiply the kg per square metre by ten to obtain the yield in tonnes per hectare.

Grass and its Conservation

Grass seed rates vary considerably, according to the type of ley or seeds mixture to be sown. The figures given in *Farm Crops* pages 128–9 can be taken as a guide.

Yields of grass can be calculated when the grass is cut for hay or silage. Even if the grass is used for grazing, it is possible to cut a few patches specially. Each one should be one square metre and from this the yield per hectare can be calculated, as follows:

$$\text{kg per m}^2 \times 10 = \text{tonnes per hectare.}$$

The following figures give some idea of the yields when grass is cut and preserved for winter feeding.

Crop	*Fresh grass*	*Silage*	*Hay*	*Dried grass*
Total yield	1 tonne	750 kg	250 kg	200 kg
Moisture content	80%	70–75%	15%	10%

FERTILISERS

The important thing about any fertiliser—whether straight or compound—is the amount of plant food it contains. The analysis figures given for any fertiliser (N, P and K) tell us two things: how it compares with any other fertiliser supplying the same plant foods, and its value for providing plant foods for our crops.

Comparing Fertilisers

As a rule, it is only straight fertilisers that have to be compared with one another to find which gives the best value. For this the cost of 1 kg of the plant food has to be found and compared with the cost of 1 kg in one or more other fertilisers. This can only be done with a group of fertilisers all supplying the same plant food—for example, two or three nitrogen fertilisers. This is the method:

(a) Find the price per tonne of each fertiliser.
(b) Divide this price by twenty to give the price per 50 kg (one bag).
(c) Find the percentage of plant food which the fertiliser contains. It is usually stated on the outside of the bag.
(d) Because the bag weighs 50 kg, the number of kilograms of N, P or K in the bag will be HALF the percentage figure quoted on the outside of the bag.
(e) Divide the price of one 50 kg bag by the number of kilograms of plant food in it to obtain the cost of one kilogram of plant food.

This can then be compared with a similar figure for any other fertiliser in the same group.

Example: We are comparing two N fertilisers and will buy the cheaper. The first costs £130 per tonne and contains 34·5% N. The other costs £100 per tonne and contains 26% N.

First

$$\frac{\pounds 130}{20} = \pounds 6{\cdot}50 \text{ per 50 kg bag;}$$

$$\frac{650 \text{ pence}}{17{\cdot}25 \text{ kg}} = 37{\cdot}7\text{p per kilo of N.}$$

Second

$$\frac{\pounds 100}{20} = \pounds 5{\cdot}00 \text{ per 50 kg bag;}$$

$$\frac{500 \text{ pence}}{13 \text{ kg}} = 38{\cdot}5\text{p per kilo of N.}$$

The first fertiliser costs 37·7p per kilo; the second 38·5p per kilo. The first is cheaper in real terms.

Use of Fertilisers

The plant food needs of any crop, in N, P and K, can be expressed as so many kilos of plant foods per hectare (or as so many units per acre). Much fertiliser is purchased in 50 kg bags (almost the same weight as one hundredweight), so first the metric system of kilos per hectare should be considered which in fact gives very similar figures to the older system of units per acre.

Remember that a 50 kg bag contains HALF the kilos of plant food as the number actually stated on the bag. If a cereal crop needs a nitrogen top-dressing of 70 kg per hectare of N, and the percentage of N in the fertiliser is shown on the bag as 34·5%, then the following calculation is needed to find out how many bags are required per hectare:

$\frac{34{\cdot}5}{2}$ divided into 70 kg per hectare gives the answer of 4 bags.

If the percentage of plant food in the bag was 26% N, then:

$\frac{26}{2}$ divided into 70 kg per hectare gives the answer of 5·4 bags.

It will also be seen from this that if a less concentrated material is used, more bags have to be handled for any given area of land. The more concentrated the fertiliser, the less has to be handled, either per hectare or per acre.

The above reasoning is simple enough when we are dealing with straight fertilisers each supplying just one plant food. However, most crops need two or three plant foods—N, P and K—at some time during their growth. It is then necessary to match the compound fertiliser to the needs of the crop, in this way.

(a) Find out the needs of the crop in kg per hectare of N, P and K.
(b) Turn this into a plant food ratio, taking the lowest number of kilograms as 1 and relating the others to it.
(c) Find which compound fertilisers are available with this plant food ratio or as near as possible. Manufacturers often print the plant food ratio on their bags.
(d) Calculate how much of these compound fertilisers will be needed per hectare to give the right amount of plant foods or as near as possible.

Example: The crop needs an application of 20 kg of N, 50 kg of P and 50 kg of K per hectare.

Take 20 kg of N as 1, then 50 kg of P and 50 kg of K are each $2\frac{1}{2}$. The plant food ratio is therefore 1: $2\frac{1}{2}$: $2\frac{1}{2}$.

On the list of available compound fertilisers there is one containing 10:23:23 and one 9:24:24. Both of these will be suitable and, remembering to halve the numbers stated on the bags, four bags per hectare should be used to apply:

either 4 × 5; 4 × 11·5; 4 × 11·5 to give 20N:46P:46K

or 4 × 4·5; 4 × 12; 4 × 12 to give 18N:48P:48K.

It is not important that the application does not match the crop requirement identically. It is close enough to be acceptable.

Some examples of plant food ratios in compound fertilisers:

Example	*Analysis*			*Plant food ratio*		
	N	*P*	*K*	*N*	*P*	*K*
Compound A	0	24	24	0	1	1
Compound B	10	23	23	1	2⅓	2⅓
Compound C	22	11	11	2	1	1
Compound D	20	14	14	1½	1	1
Compound E	17	17	17	1	1	1
Compound F	15	15	21	1	1	1½
Compound G	17	8	24	2	1	3
Compound H	12	18	12	1	1½	1

For units per acre, the method of matching the plant food ratios of the crop and the selected fertiliser still applies. But in order to find the application rate per acre use the whole number stated on the bag, *do not halve it* in this case.

A number of farmers still prefer to consider the application of fertilisers as units per acre, and there is a very simple conversion for this:

Kilograms per hectare × 0·8 = units per acre,

or Units per acre × 1·2 = kilograms per hectare.

Pallet Handling

Fertilisers (and for that matter other items) can be delivered on pallets. Using fork-lift equipment saves a lot of time in unloading delivery lorries and also when fertilisers are being loaded on to trailers or distributors. Each pallet normally holds 1·5 tonnes and takes up the space of 1·5 m × 1·2 m and 1·0 m height.

For safety reasons, loaded pallets are not usually stacked more than three high. For a given size of building therefore, it is possible to calculate how many pallets of fertiliser can be stored under cover. Quite often a lower price per tonne is quoted for fertiliser on pallets and a pallet hire charge is made which is credited when the empty pallet is returned.

Example: If 165 tonnes of fertiliser are ordered on pallets each year and the hire of a pallet is £9·00, what floor area must be found to stack the fertiliser and what is the hire charge?

$$\frac{165}{1 \cdot 5} \text{ tonnes} = 110 \text{ pallets}$$

$$\frac{110 \text{ pallets}}{3 \text{ high}} = 37 \text{ stacks}$$

$$37 \text{ stacks} \times 1 \cdot 5 \text{ m} \times 1 \cdot 2 \text{ m} = 67 \text{ m}^2 \text{ of floor area.}$$

The hire charge will be 110 × 9 = £990.

Use of Slurry

Because livestock are kept more intensively and the cost of straw can be expensive in areas where cereals are not grown much, animal residues are frequently handled and spread as slurry. Some farmers regard this as a chore and are glad to dispose of it on any spare land. It does, however, contain valuable nutrients available as a partial alternative to purchased fertilisers.

The nutrient content will vary with the amount of dilution by water and the type of livestock but a typical analysis would be:

		Kg of nutrient per cubic metre		
		N	*P*	*K*
Cattle:	10% dry matter	2·5	1·0	4·5
Pigs:	10% dry matter	4·0	2·0	2·7
Poultry:	25% dry matter	9·1	5·5	5·4

If it is applied to the land by slurry tanker, provided the capacity of the tanker is known and the number of full loads spread is recorded it will be possible to calculate some estimate of the nutrients applied and the saving to be made on compound fertilisers.

Example: A slurry tanker holding 6,750 litres spread eighty loads of average cow slurry on a 15-hectare field. How many kilograms per hectare of N, P and K were assumed to be applied?

Tanker capacity of 6,750 litres ÷ 1,000 = 6·75 m^3 per load.

$$\frac{6{\cdot}75 \times 80 \text{ loads}}{15 \text{ ha}} = 36 \text{ m}^3 \text{ per ha.}$$

Therefore:

$$36 \times 2{\cdot}5 = 90 \text{ kg of N per ha,}$$

$$36 \times 1{\cdot}0 = 36 \text{ kg of P per ha,}$$

$$36 \times 4{\cdot}5 = 162 \text{ kg of K per ha.}$$

Lime

The acidity of the soil is measured in terms of pH (see *Farm Crops* page 56). As the pH figure drops below 7, the soil becomes more and more acid, and it becomes necessary to add lime to the soil to correct this.

As a rough guide, to raise the pH by 0·5 the application of 2·5 tonnes of burnt lime per hectare is needed, or its equivalent in some other type of lime (see table on page 39).

Example: To raise the pH of a medium loam from 5·5 to 6·5, 5 tonnes of burnt lime per hectare must be applied.

Unit Cost

There are several different forms of lime which can be used. As for all bulky material, transport costs and spreading costs are important and must be taken into account in any calculations. For this reason, get costs of lime delivered and spread on the land.

Each type of liming material has a neutralising value which

shows how it compares with other materials; using this, you can calculate the value of any liming material in relation to any other, and decide which to use. This is how to do it:

(a) Find the current cost of the liming material per tonne delivered and spread on the land.
(b) Divide this figure by the neutralising value of the material, as shown in the table below. This gives a Unit Value.
(c) Compare the unit values of the different liming materials which you can get locally.

Example: You can buy ground limestone delivered and spread at £12·00 per tonne. It has a neutralising value of 50:

$$\frac{£12{\cdot}00}{50} = \frac{1{,}200\text{p}}{50} = 24 \text{ pence per unit.}$$

Liming material	*Neutralising value*	*Quantity needed to give the same liming effect*
Burnt lime	100	1 tonne
Slaked (hydrated) lime	75	$1\frac{1}{3}$ tonnes
Ground limestone or chalk	50	2 tonnes
Lump limestone or chalk	33	3 tonnes
Waste factory lime	25	4 tonnes

(These values, particularly for lump limestone and factory lime, are given here in a very simple form; they need to be checked locally as this type of material varies in value.)

Spray Chemicals

It is possible to compare different spray chemicals in the same way as fertilisers, taking into account their real value—the amount of active ingredient which each one contains. There are many firms making these chemicals, and hundreds of sprays under different trade names.

The active ingredients are stated either as a percentage by weight in the case of the powders, or as a percentage by volume in the case of liquids. In practice, the best way to compare them is to work out the cost of recommendations as to the quantities per hectare.

Example: Product A costs £53 per 10 litres and is used at the rate of 4·3 litres per hectare.

Product B costs £79 per 20 litres and is used at the rate of 5·7 litres per hectare

A $\frac{£53}{10} = £5{\cdot}30$ per litre $\times 4{\cdot}3 = £22{\cdot}79$ per hectare.

B $\frac{£79}{20} = £3{\cdot}95$ per litre $\times 5{\cdot}7 = £22{\cdot}51$ per hectare.

There is very little difference, although Product B is cheaper by 28p per hectare.

The purchase of spray chemicals is common practice on most farms. It is important to ensure that sufficient spray is bought without too much left over for the following season. A spray order is prepared in this way:

(a) A decision is first made on the choice of weedkillers, fungicides and insecticides likely to be used on different crops on the farm.
(b) You then find out the application rate in litres (for liquids) or kilograms (for solids) per hectare and also the pack size in which they are delivered.
(c) Multiply the application rate per hectare by the number of hectares of crop on which each product will be used and divide by the total amount in each pack.
(d) Round up this total to the nearest whole number and subtract the number of packs remaining in store from the previous season to give the number which should now be ordered.

Example: A weedkiller is applied at 3·5 litres per hectare, and 60 hectares of cereals are to be sprayed. The cans contain 23 litres each and two cans remain in store. How many more cans should be ordered?

$$\frac{3{\cdot}5 \times 60}{23} = 9{\cdot}1\text{; therefore ten cans are required.}$$

Two cans remain in store so eight cans should be ordered.

This routine can also be followed when buying additives for silage making and preparing fertiliser orders.

Crop Sales

Most sales of crop products are on a fairly straightforward basis. The price is known or agreed at so much a tonne or multiples of a kilogram and the total price can be calculated simply.

Example 1
84 tonnes of potatoes are in store and these are being sold at £70 per tonne. This sale will bring in a total of 84 × 70 = £5,880.

But if half of these are sold at the farm gate at £2·50 per 25 kg and the rest at £70 per tonne, then the increased income would be:

	£
42 tonnes × 40 (bags per tonne) × £2·50	= 4,200
42 tonnes × £70	= 2,940
	£7,140

Example 2
There are 55 tonnes of wheat at £128 per tonne and 30 tonnes of barley at £120 per tonne left in store. The total value would be:

	£
55 × £128	= 7,040
30 × £120	= 3,600
	£10,640

Example 3
Rye-grass seed is being sold at £28·00 per 50 kg and we have 3·4 tonnes to sell:

£28 × 20 = £560 per tonne

£560 × 3·4 = £1,904 income.

Sale by Quality

Many prices these days are influenced by the quality of the product. Above the basic price there are premiums available if the crop offered for sale can be used as human food, has the correct moisture or sugar content and/or is free from impurities and contamination by weed seeds, disease and insect damage.

It should also be remembered that crop products not reaching a certain standard may incur price penalties and their value will be lower than the basic price on offer.

Calculations of returns must take into account these premiums and deductions if they apply.

As a deterrent to over-production a levy is now imposed on all cereals sold off the farm. Currently in 1989 it is £7·32 per tonne and all or part of this will be deducted by the buyer from the quoted price. The net price is what the farmer will receive.

Example: If a good breadmaking wheat yields 6·0 tonnes per hectare and commands a price of £130 per tonne, what yield of feed wheat must be grown, valued at £122 per tonne, to give the same return per hectare?

$$6{\cdot}0 \times (£130{\cdot}00 - £7{\cdot}32) = £736{\cdot}08$$
$$£746{\cdot}08 \div (£122{\cdot}00 - £7{\cdot}32) = 6{\cdot}42 \text{ (tonnes per hectare)}$$

This is the yield of feed wheat which must be grown to bring the same return.

FARM CROP PROBLEMS

1. The area of a triangle is 4·2 m^2 and the base is 3 m. What is the height?
2. What length of electric fence, to the nearest metre, is needed to cover the longest side of a rectangular field of 2 ha? The short side measures 110 m.
3. In a rectangular field of 6·5 ha, a row crop is being drilled in rows 0·45 m apart parallel to the longest side which measures 373 m. How many rows of the crop will there be?
4. How many parallel tile drains can be laid in a field which measures 152 m across if they are laid at 8 m intervals?
5. We are putting up a cylindrical grain storage bin in a barn. The diameter is 5·5 m. What area of floor space will it cover and what will be the circumference?
6. A fuel tank in the yard measures 2·4 m by 1·5 m by 1·5 m. What is its cubic capacity and how many litres of diesel fuel will it hold?
7. The cylindrical tank of a low-volume sprayer has a diameter of 0·75 m

and is 1·3 m long. How many litres of water will it hold? If the sprayer weighs 288 kg, what will be the total weight when full?

8. What quantities to the nearest tonne of wheat, barley and oats can be stored in a grain silo measuring 3 m by 3·65 m by 4·88 m high?
9. For the information given in the table on page 28, work out the kg per cubic metre for hay, all the straws and silage.
10. How many tonnes of silage at 25% dry matter will there be in a clamp 13·5 m wide, 33·5 m long and 1·85 m high?
11. In a particularly wet season, a farmer makes silage of 21% dry matter. How many tonnes of material will then be stored in the clamp referred to in question No. 10? For a clamp of this size, compare the total amount of dry matter which can be stored of silages at 21% and 27% dry matter.
12. Would an area 8·6 m square and 2·45 m high be large enough to store 6 ha of maincrop potatoes yielding 30 tonne per hectare?
13. A clamp of potatoes inside the earth and straw covering measures 1·38 m high, 1·85 m at the base and 30 m long. What weight of potatoes is there to the nearest tonne?
14. What total storage capacity is needed to store grain from 32·4 ha of wheat, 44·5 ha of barley and 10 ha of oats assuming yields stated in the chapter? How many circular silos of 5·5 m diameter and 5 m high must we install if we want to avoid mixing any of the cereals together?
15. The Thousand Grain Weight of a winter wheat variety is declared as being 46 grams. How much seed must be ordered to drill 55 ha of this variety? A second variety has a T.G.W. of 41 gm. By how much does the order differ if this variety is chosen for drilling instead?
16. The seed rate for winter barley achieved only 370 seeds drilled per square metre. Assuming normal losses of 12·5% for germination and post drilling followed by a harsh winter giving 20% winter kill, how many plants per square metre are likely to survive through to the spring?
17. In the autumn, a farmer took samples of kale from eight different sites in the field. If the kale was drilled in rows of 0·45 m, what length of row did he need to cut to obtain the weight per square metre? If his sample weighings in kg were 4·2; 5·4; 5·2; 5·6; 6·3; 4·8; 5·1; 5·9, what was the average yield in tonnes per hectare?
18. Suggested plant food applications in kg per ha for certain arable crops are winter wheat 0:50:50 in the autumn and two top dressings of 50N and 125N in the spring; spring barley 90:45:45; sugar beet 130:60:180; maincrop potatoes 190:190:265; grass for first-cut silage 85N and four weeks later 50.50.70. Using information given in the chapter, choose the correct fertiliser for each crop and work out the amounts for each crop. Also calculate the amounts of fertiliser which should be applied per hectare.

19. Compare the cost of a kilogram of plant food in (a) superphosphate containing 18% P costing £70 per tonne and (b) triple superphosphate containing 45% P costing £130 per tonne.

20. A barn 12 m wide by 16·5 m long and 4·4 m high is available for storing fertiliser on pallets. Ignoring any space which must be left for access, how many tonnes can be put under cover?

21. A farmer spreads 17 × 50 kg of 34·5% N fertiliser and 7 × 50 kg of a 40% P fertiliser on each hectare of his grassland. How many kg of N and P are applied per hectare?

22. The slurry from a pig unit amounts to 640,000 litres per year and this is applied to a 25-ha field prior to potato planting, using a slurry tanker which holds 12 m^3. How many kg of each plant food are applied per hectare and how many journeys are made to the field?

23. The pH of a field is 5·3. How much (a) hydrated lime and (b) ground limestone should be applied per hectare to raise the pH to 6·5?

24. Two similar sprays are available. Product A costs £45·00 for 4 litres and is used at 1 litre per ha. Product B costs £24·00 for 1 litre and is used at 0·5 litre per ha. Which is the cheaper material to use?

25. A growth regulator costs £19.35 per ha when applied at 1·5 litres per ha. What will be the cost of a 10 litre can?

26. How many cans of spray chemical should be ordered for 60 ha of barley if the rate of application of the herbicide is 3·5 litres per ha and it is sold in 20 litre cans, and the fungicide is applied twice at 0·5 litres per ha and sold in 5 litre cans?

27. How much will a farmer receive for 36 tonnes of barley delivered at 18% moisture but valued at £110 per tonne at 16% moisture? The drying losses are 6·5% weight and the merchant charges £4·25 per tonne per 1% removed of moisture. The levy is £7·32 per tonne on the grain finally sold.

28. The NIAB leaflet shows that cereal variety A is a malting variety of barley with a potential yield of 98 and variety B is a feeding variety rated at 107 when both are compared to the average of 100. If the malting price on offer is £130 per tonne and the feed price is £108 per tonne, how much more must variety B yield compared to: (i) variety A, and (ii) its own potential, to bring in the same income per hectare?

29. Each irrigation sprinkler covers 0·02 ha. How many sprinklers are needed to cover (a) 0·5 hectare (b) 2·8 hectares?

30. To apply 25 mm of rain on 1 ha needs 254 m^3 per ha. How much water in litres is needed to apply (a) 125 mm on 2·5 ha (b) 75 mm of water on 18 ha?

Chapter 3

Farm Livestock Calculations

Chapter 3

Farm Livestock Calculations

LIVESTOCK FEEDS

Livestock Units

TO GET some idea of how much grazing is needed for the cattle and sheep on a farm, or roughly to work out quantities of bulky foods needed over a certain period, the figures below can be used. They compare the food needs of sheep, young cattle and horses with those of dairy cows.

Class of livestock		*Livestock Units per head*
Dairy cow		1·0
Bullock		0·8
Young stock	18–24 months	0·7
	12–18 months	0·6
	6–12 months	0·4
Lowland ewe (including lambs)		0·2
Ram and teg		0·2
Hill ewe		0·1
Horse		1·0

Grazing

Just how much stock can be carried on an area of grassland depends on the type of land and the quality of the grass. The second table gives some idea of how different types of grassland compare.

The Grazing Livestock Units used here are not the same as the Livestock Units above. A grazing livestock unit is something like the number of dairy cows (or their equivalents from the above table) for which one hectare of the grassland will provide grazing and grassland conservation for one complete year.

Type of grassland	*Grazing Livestock Units per forage hectare*
Very best grassland	2·5
Excellent permanent pasture and new leys	2·05
Good permanent pasture and older leys	1·75
Poorer grassland	1·45

The levels of fertiliser application throughout the summer period will clearly affect these units, sending them either up or down.

Rationing Food by Area

When livestock are grazing in a grass field, or folded on kale or roots, they are rationed by area. Often this is done by eye or by experience. The farmer can judge from the state of the grass, and how the cattle or sheep are doing, just how much food they need and how much they are getting. Sometimes it is necessary to calculate how much grazing a field will provide and what amount should be allowed per head.

Example: A common allowance for cows strip grazing on grass in the spring is 100 cows per hectare (or 10,000 m^2) per day.

A grass field 250 metres wide requires a strip

$$\frac{10{,}000}{250} = 40 \text{ metres deep each day (or one move of 20 metres after each milking)}$$

to give a hectare of grazing.

To assess the quantity of grass available for each cow, cut a few sample patches of 1 square metre each and weigh them. If they average 0·6 kg per square metre each cow will be getting 60 kg of grass daily, which is a reasonable amount.

Using the simple formula that kg per $m^2 \times 10$ is equal to tonnes per hectare, in this case the actual yield of grass is about 6·0 tonnes per hectare.

Rationing Food by Volume

Sometimes cattle and sheep are allowed to help themselves to food—this is known as *ad lib* feeding. When it is done with silage, it is known as the 'self feed system'. We can estimate the quantity they should have per head per day, an average amount over the herd, and ration them accordingly.

This is done either by cutting the right quantity and throwing it to them, or allowing them to feed through an adjustable barrier.

Example: A clamp of silage in a building is 2 metres high and 14 metres wide; 80 cows are feeding on this silage. In practice we find they can pull it out from the surface 0·15 metres (15 cm) deep. First find out the width each cow will be getting:

$$\frac{14}{80} = 0{\cdot}175 \text{ metres width per cow.}$$

Now we know that each cow will get, on average per day, 2 metres height by 0·175 metres width by 0·15 metres depth. Working this out we get $2{\cdot}0 \times 0{\cdot}175 \times 0{\cdot}15 = 0{\cdot}05\ m^3$. So each cow gets $0{\cdot}05\ m^3$ per day. We know that $1{\cdot}5\ m^3 = 1{,}000$ kg (see page 28).

therefore $$1{\cdot}0\ m^3 = \frac{1{,}000}{1{\cdot}5} = 667 \text{ kg}$$

and $$\begin{aligned} 0{\cdot}05\ m^3 &= 667 \text{ kg} \times 0{\cdot}05 \\ &= 33 \text{ kg of silage per cow per day.} \end{aligned}$$

Rationing Cattle and Sheep

Working out properly balanced rations for livestock can be a complicated job if we use analysis figures for Metabolisable Energy and Digestible Crude Protein. There are simpler methods which give quite reasonable results in practice—and these rations can always be checked by more accurate methods if necessary.

One simple method, based on the Hay Equivalent System, is set out on page 48. This shows the amount of hay (other bulky foods can be partly used in place of hay) to be allowed for each animal, daily, to cover *Maintenance* needs. Quality of hay varies, and these figures are based on medium-quality hay.

Type of stock	Hay requirement for maintenance needs
Dairy cows	1 kg of hay per day for each 66 kg of liveweight Friesians 600 kg — 9 kg of hay Ayrshire 500 kg — 7·5 kg of hay Guernsey 450 kg — 6·8 kg of hay Jersey 380 kg — 6·0 kg of hay
Store beef cattle	1·5 kg of hay per day for each 50 kg liveweight
Finishing beef cattle	1·25 kg of hay per day for each 50 kg liveweight
Calves 3–12 months	0·5 kg of hay per day for each month of age
Youngstock over 12 months	0·3 kg of hay per day for each month of age
Sheep	1·0 kg of hay per week for each 5·0 kg of liveweight

Examples:
100 Friesians would need 900 kg of hay per day.
An 8-month-old calf needs 4·0 kg of hay per day.
A 70 kg sheep needs 14 kg hay per week, about 2 kg per day.

Using other Foods

When you have worked out the amount of hay needed by an animal, you can substitute for some of it quantities of other foods, according to what is available on the farm. Use these figures:

For 1 kilogram of medium-quality hay:	
	0·6 kg barley, oats or sugar-beet nuts
	0·75 kg dried grass or excellent hay
	3·0 kg grass silage, potatoes or fodder beet
	4·0 kg kale, beet tops or swedes
	5·0 kg mangels or wet beet pulp
	7·0 kg turnips

Example: A Friesian cow would need 9 kg of medium hay per day for maintenance. We only have enough hay to allow 3 kg per day and must also use grass silage and rolled barley.

5 kg of hay is equal to $5 \times 3 = 15$ kg grass silage;
1 kg of hay is equal to $1 \times 0{\cdot}6 = 0{\cdot}6$ kg rolled barley.

Thus the ration for the cow could be 3 kg of hay, 15 kg of grass silage and 0·6 kg of rolled barley.

Rationing Pigs

As pigs are usually fed on meal or nuts rather than a mixture of bulky foods and concentrates like cattle and sheep, it is simpler to calculate their food requirements.

Using the figures in the table below, you can work out the needs for a herd of any size for a year, or for a shorter period.

		Meal or nuts
Boar	One year, all requirements	0·75 tonne
Sow	One year, all requirements (including creep feed for litters)	1·15 tonnes
Pork pig	Weaning to 68 kg liveweight	0·14 tonne
Bacon pig	Weaning to 90 kg liveweight	0·24 tonne
Heavy hog	Weaning to 118 kg liveweight	0·38 tonne

Example: A breeding herd of 50 sows and 3 boars would need in a year:

$$50 \times 1{\cdot}15 = 57{\cdot}50 \text{ tonnes}$$
$$3 \times 0{\cdot}75 = 2{\cdot}25 \text{ tonnes}$$

Total 59·75 tonnes

Individual rations can be worked out using figures given in *Farm Livestock.*

Examples: A suckling sow should have 2·3 kg food per day for herself and 0·34 kg for each piglet in the litter. So a sow with ten piglets should get $0{\cdot}34 \times 10 = 3{\cdot}4$ kg for the piglets and 2·3 kg for the sow, a total of 5·7 kg per day.

A growing pig for finishing to bacon should get 0·1 kg of food per week of age for every day in that week (up to a maximum quantity which varies with size and type). So an 18-week-old baconer just prior to slaughter should get $0{\cdot}1 \times 18 = 1{\cdot}8$ kg of food per day.

LIVESTOCK YIELDS

Milk Yield

There are so many things that influence milk yield that it is difficult to tell in advance the total amount of milk a cow will give in a lactation. This is even more difficult when no proper records are kept. However, if the 'peak yield' is known the total yield in the lactation can also be estimated. Peak yield is the daily yield when the cow is producing her highest daily quantity of milk, usually about six weeks after calving. To find the total lactation yield from this:

For heifers:	multiply peak yield by 220
For cows:	multiply peak yield by 200.

Example: A heifer gives a peak yield of 21 litres. Thus her total lactation yield will be about 220 × 21 = 4,620 litres.

Milk Quality

Milk quality is now very important in profitable milk production—price depends on quality. There are differences between breeds, between herds and between individual cows in a herd. Moreover, there are differing standards of hygiene in the production of milk.

For both compositional quality and hygienic quality the milk from each farm is centrally tested each month by the Milk Marketing Board and an average result is declared to the farmer. It is on this basis that his month's supply of milk is paid.

The milk is tested for the:

- percentage of butterfat
- percentage of protein
- percentage of lactose
- Total Bacterial Count (TBC) per ml.

For each of the three compositional constituents of milk, the average monthly result is multiplied by an appropriate price in pence per 1%. The three subtotals are added together to fix the basic price per litre to be paid to the farmer for that month's supply.

In addition, there is an extra payment or deduction for the hygienic quality according to the following chart.

	Total Bacterial Count per ml	*Addition or deduction*
Band A	up to 20,000	+0·23p per litre
Band B	21 to 100,000	no addition or deduction
Band C	more than 100,000	−1·5p if no deductions in past 6 months −6·0p if 2·4p deduction has been made in past 6 months −10·0p if 3·6+p deduction has been made in past 6 months

Example: A dairy farmer is notified that the average composition of the milk supplied from his farm for the month of April is 4·24% butterfat, 3·31% protein and 4·75% lactose. The rate of payment for 1% of each constituent is: 2·174p for butterfat, 2·217p for protein and 0·327p for lactose*. The Total Bacterial Count was 18,000. What will be the price per litre of milk that he will receive?

			pence
Butterfat	2·174p × 4·24	=	9·218
Protein	2·217p × 3·31	=	7·338
Lactose	0·327p × 4·75	=	1·553
TBC	Band A +0·23p	=	0·230
			18·339

The price in pence per litre for April's milk would be 18·339p.

The MMB makes a further addition or deduction overall to the farmer's milk price depending on the month of the year. This seasonal variation is aimed at encouraging more milk when supplies are low—that is when a lot of cows are dry. A price deduction is applied when there is a large milk surplus, normally in the spring when cows are turned out to grass. You should look in the MMB publication, *Milk Producer*, to find out what these extra additions or deductions are.

Sow Yields

A sow can produce two litters in a year, with ten pigs or more in each litter. Many sows do not do as well as this, either producing

*These values apply for April and May 1989 but the principle of calculation remains the same if the values change.

fewer pigs, or taking longer than twelve months to produce two litters. There are also many losses of young pigs between birth and weaning, due to disease and bad management.

Example: One sow produces one litter of eight and another litter of twelve, within twelve months. Her yearly production is twenty pigs.

Another sow produces one litter of ten and a second litter of eleven within fourteen months. Her total production in this time is twenty-one pigs, but this is not her yearly production, which must be calculated this way:

$$\frac{21}{1} \times \frac{12}{14} = \frac{3 \times 12}{2} = 18.$$

Her yearly production is eighteen pigs.

Sows which have two litters per year take 365 ÷ 2 = 182 days to complete one cycle from service to service. If pregnancy lasts 115 days then the suckling/empty sow period is 67 days of which 56 days might be the suckling period. With earlier weaning the yearly production of a sow can be increased.

Example: If three-week weaning is being practised on a pig unit, how many litters could a sow produce in one year?

One cycle would be 115 days pregnancy + 21 days suckling + 7 days empty prior to service = 143 days.

$$\frac{365}{143} = 2{\cdot}55 \text{ litters per sow per year.}$$

But every time a sow returns to service another 21 days of production has been lost. For each return to service, subtract 21 days from the total days in one year and then divide by the number of days of one cycle (pregnancy + suckling + empty) practised on the farm.

Example: In a three-week weaning pig unit a sow returns to service on one occasion during the year (365 days).

$$\frac{365 - 21}{143} = 2{\cdot}4 \text{ litters per year will be the most this sow can achieve in that year.}$$

Ewe Yields

The production of a flock of breeding ewes is usually shown as a 'lambing percentage'. The production of a single ewe is usually given as a simple fraction.

If 100 ewes produce 150 lambs between them, we say that the flock has a lambing percentage of 150%, and one of these ewes produces 'a lamb and a half'.

To take into account any lambs which die and ewes which die or are barren, compare the number of lambs reared with the number of ewes *put to the ram* in the autumn. This gives a more realistic figure.

Example: A flock of 240 ewes put to the ram produces 420 lambs reared. The lambing percentage will be:

$$\frac{420}{240} \times \frac{100}{1} = \frac{4,200}{24} = 175\%.$$

GROWTH RATE

Dairy

For autumn-calving herds, the aim is to bring heifers into the milking herd for the first time in the autumn. If they themselves are autumn-born this will mean that they will be either two years or three years old when they can calve.

Most dairy farmers try to keep the age of a down calving heifer as young as possible provided she has become a well-grown animal by the time she calves. The table below gives an idea of the growth of heifers of different breeds aiming to calve at two years. If they are to calve later then the growth rate can be reduced to achieve the same body weight at calving.

	Weight of animal (kg)			*Increase in body weight (kg)*
Breed	*Birth*	*12 weeks*	*Calving*	
Jersey	27	66	345	318
Guernsey	29	75	370	341
Ayrshire	32	80	400	368
Friesian	43	84	510	467

From these figures the average daily or weekly rate of growth can be calculated.

Example: The Ayrshire increases by 368 kg liveweight in 104 weeks. It therefore grows at an average rate of:

$$\frac{368}{104} = 3{\cdot}54 \text{ kg per week}$$

or

$$\frac{368}{728} = 0{\cdot}51 \text{ kg per day.}$$

Beef

Young stock for beef production have to grow faster than dairy heifers. Under modern intensive systems they have to grow very fast. The table gives an idea of the rate of growth under different systems.

Weight of animal (kg)		*Age at slaughter*	*System of production*
Birth	*Slaughter*		
45	530	2 years	Extensive
45	430	1 year	Intensive

From these figures the average daily or weekly rate of growth can be calculated.

Example: Under the extensive system, the animal puts on 485 kg in a hundred weeks. It therefore grows at an average rate of:

$$\frac{485}{100} = 4{\cdot}85 \text{ kg per week}$$

or

$$\frac{4{\cdot}85}{7} = 0{\cdot}69 \text{ kg per day.}$$

Under the intensive system, the animal puts on 385 kg in about fifty weeks. It grows at an average rate of:

$$\frac{385}{50} = 7{\cdot}7 \text{ kg per week}$$

or $$\frac{7 \cdot 7}{7} = 1 \cdot 1 \text{ kg per day.}$$

Pigs

The growth rate of pigs depends on the breed and type of the pigs, and the way they are fed and managed. Pigs will grow faster if they are fed more food—but if they have too much they grade badly because they get too fat, and thus make a lower price per kg. Or they may eat too much food in relation to their growth, and so cost too much. The table below shows rates of growth.

Type of pig	*Weight of pig (kg)*		*Period of fattening*
	Weaning at 5 weeks	*Slaughter*	
Porker	10	60	14 weeks
Baconer	10	90	19 weeks
Heavy hog	10	120	24 weeks

Example: The porker puts on 50 kg liveweight in fourteen weeks. It grows at the rate of 50 ÷ 14 = 3·6 kg per week, or 0·51 kg per day.

Killing-out Percentage

The killing-out percentage shows the proportion of the animal's carcase which can be sold by the butcher. The deadweight is given as a percentage of the weight of the live animal. It differs between different classes of animals—pigs, sheep and beef cattle—and between ages of animals. The younger the animal, the lower is its killing-out percentage, because it has more bones and intestines in relation to its whole weight.

Example: A bullock slaughtered at 490 kg liveweight kills out at 53%. Its deadweight is:

$$\frac{53}{100} \times \frac{490}{1} = \frac{25{,}970}{100} = 259 \cdot 7 \text{ kg deadweight.}$$

Cattle and sheep have a low killing-out percentage, because their digestive systems are so big and their heads are not included in the carcase weight. Pigs have a much higher killing-out percentage.

Example: A bacon pig of 89 kg liveweight kills out at 75%. Its deadweight is:

$$\frac{75}{100} \times \frac{89}{1} = 66{\cdot}8 \text{ kg deadweight.}$$

The table below shows average weights, both live and dead, and killing-out percentages, of various types of animal.

	Liveweight (kg)	*Killing-out percentage*	*Deadweight (kg)*
Porker	60	70	42
Baconer	90	75	67·5
Heavy hog	118	76	90
Finished lamb	34	50	17
Cull ewe	72	55	40
Bullock	520	54	281

Food Conversion

Sometimes we have to think of farm animals as machines which are used to turn feedingstuffs into something else—meat, milk or eggs. Some of these 'machines' are better at the job than others. We measure their efficiency by calculating the food conversion rate.

This can be done quite easily with pigs which eat one food all the time—some sort of meal. It is more complicated with cattle or sheep, which are fed on a mixture of feeds, some wet and some dry, but it can be done by working out the ration in terms of 'dry food'.

This is the way to calculate any conversion rate:

(a) Find out the liveweight gain of an animal within a certain period, or its production of milk or eggs, in kg. For liveweight gain, deduct the starting weight from the finished weight.
(b) Find out the total quantity of food the animal has eaten during this period—either as kg of meal or kg of dry food.

(c) Divide the kg of food by the kg of liveweight gain (or kg of other form of production).
(d) The final figure shows how many kg of food it takes to produce 1 kg of liveweight gain (or other form of production).

Example: A weaner weighs 15 kg and increases in liveweight to 90 kg by the time it goes to the bacon factory. It has gained 75 kg liveweight during this period.

To make this extra weight, it has eaten 263 kg of meal. Its conversion rate is found by this calculation:

$$\frac{263}{75} = 3{\cdot}5.$$

It has taken 3·5 kg of meal for the pig to produce 1 kg of liveweight gain; its conversion rate is 3·5 to 1.

If it had only eaten 225 kg of meal, its conversion rate would be:

$$\frac{225}{75} = 3, \text{ that is } 3{\cdot}0 \text{ to } 1.$$

Costs

Another useful exercise is to cost the food eaten and find from this the cost of each 1 kg of liveweight gain (or eggs or milk). Compare this cost with the price received for the product.

Remember that 1p per kilogram = £10 per tonne. Using this simple formula, it is easy to calculate the price of a kilogram of feedingstuffs; for instance:

£150 per tonne = 15p per kg, and
£185 per tonne = 18·5p per kg.

Example: The rearer meal fed to the pig in the example above costs £199·00 per tonne which is 19·9p per kg.

If the conversion rate is 3·5, it costs 3·5 × 19·9p = 69·65p to produce 1 kg of liveweight gain.

If the conversion rate is 3·0, it costs 3·0 × 19·9p = 59·7p which is a saving of 9·95p per kg of liveweight gain.

LIVESTOCK SALES

Sales by Liveweight

Livestock can be sold either at so much per animal, or on the basis of weight and quality. For example, weaner pigs can be sold in the market at whatever price they will fetch per head.

On the other hand, some buyers prefer to go to farms and buy the pigs they want at a price usually related to a weight per weaner of 25 kg.

Similarly, finished cattle, pigs and lambs can be taken to market, passed across a weighbridge, graded, and then bid for in the auction ring.

For cattle and pigs, prices are quoted at pence per kilogram liveweight. For finished lambs, an estimate is made of the dead carcase weight (see section on killing-out percentage) and prices are quoted in pence per kilogram of estimated dead carcase weight (edcw).

Currently three examples would be:

(a) 510 kg steer at 116 pence per kg = £591·60 per animal
(b) 75 kg cutter pig at 90 pence per kg = £67.50 per animal
(c) 16 kg (edcw) lambs at 210 pence per kg = £33.60 per animal

Sales by Deadweight

A reliable method of selling livestock is on a deadweight price, as this benefits both buyer and seller. The price is based on the actual deadweight of the animal. These prices are quoted regularly in the agricultural magazines. To calculate what an animal is worth in this way, use this method:

(a) Get the deadweight of the animal, if this is known. If not, calculate the deadweight from the liveweight (see page 55).
(b) Take the current prices, which include any subsidy payments.
(c) Multiply the deadweight of the animal by the price in pence per kilogram. This will give the total price for the animal.

Example: A bullock of 480 kg will have a killing-out percentage of 56%. Its deadweight will be:

$$480 \times \frac{56}{100} = 268{\cdot}8 \text{ kg.}$$

If the price for a beast of this type is 210 pence per kilogram deadweight then the value of the animal will be:

$$268{\cdot}8 \times £2{\cdot}10 = £564{\cdot}48$$

Conversion

When you look up prices for fatstock, make sure whether they are given for liveweight or deadweight; sometimes this can be confusing. You can, however, work out your own comparison between liveweight and deadweight prices but it will be necessary to use an average killing-out percentage such as those quoted on page 56.

The calculation is as follows:

either $$\text{pence per kg liveweight} \times \frac{100}{\text{killing-out }\%} = \text{pence per kg deadweight,}$$

or $$\text{pence per kg deadweight} \times \frac{\text{killing-out }\%}{100} = \text{pence per kg liveweight.}$$

Example 1: A bacon pig was paid for on contract at 114p per kg deadweight. What would be the equivalent price if sold liveweight through an auction market?

$$114 \times \frac{75}{100} \ 85{\cdot}5\text{p per kg liveweight.}$$

Example 2: A bacon pig kills out at 69 kg deadweight. The price for a top-grade baconer (Grade 1) of this weight on contract is 114p per kg. For a Grade 2 baconer the price received is 107p per kg. Therefore the price received for each pig according to grade will be:

Grade 1	69 × 114p = £78·66
Grade 2	69 × 107p = £73.83.

The difference in value between baconers of these two grades is £4·83.

FARM LIVESTOCK PROBLEMS

1. The livestock on a farm consists of 150 cows, 30 in-calf heifers, 40 heifers aged 12–15 months and 40 six-month-old calves, together with 300 lowland ewes and 8 rams. How many livestock units does this represent?
2. How many bullocks weighing 400 kg can be grazed on 12 hectares of new leys and 7·5 hectares of poor grassland in one season?
3. How much of the very best grazing must be allowed for a herd of 120 Friesian cows in one season?
4. An 80-cow herd is to be strip grazed across a field which measures 275 metres on its longest side. How far must the electric fence be moved each day to give each cow a reasonable amount of grass? If the field is 5·25 hectares, how many days will the grazing last?
5. If a Friesian cow can eat about 45 kg of silage per day on a self-feed system, how much silage is being eaten by a 75-cow herd: (a) per week; (b) per month of 31 days; (c) in a 200-day winter?
6. How many days will a silage clamp of 750 tonnes at 25% dry matter provide enough feed (self-fed) for a herd of 90 Friesian cows which can eat 45 kg per day?
7. If a silage clamp is 2 metres high and each cow is allowed 150 mm width at the feeding face, how far back must the feed barrier be moved each day to allow 30 kg of 25% silage per cow?
8. How many kg of average hay will a herd of 75 Ayrshires need each day? By how much is this total reduced for a similar number of Jerseys?
9. A flock of half-bred sheep comprising 320 ewes and 8 Suffolk rams (average liveweight 70 kg each) will need hay feeding for the last fortnight in January and all of February. How much hay must be available for this period?
10. A Friesian cow will get 5 kg of hay and 8·5 kg of silage per day. How much sugar beet nuts does she need daily to complete the maintenance ration?
11. If only 5·5 kg of hay is available daily for each Friesian cow in a herd, what quantities of each of these other foods would be needed to provide a complete maintenance ration; silage, kale, mangels, barley?

12. What total quantities of hay and silage are needed for 60 Guernsey cows if each is to get 2·5 kg hay and 13 kg of silage per day for the six winter months?
13. A farmer with 4 boars and 70 sows rearing 20 piglets per sow per year which themselves are taken to bacon weight needs to know what his food requirements will be for the year so that he can contract forward with his merchant. What will be the total amount?
14. In allocating pig food to the farrowing house for one week, the farmer calculates that there will be 22 sows and 198 piglets. How much sow and weaner food will be consumed during this time?
15. A bacon house has a throughput of 750 pigs a year from weaning to slaughter. How many tonnes of food will be eaten by these pigs in a year? If home-grown barley forms 85% of the ration, how many tonnes of feeding barley must the farmer retain for his own use?
16. In a milking herd of 75 cows, the 16 heifers peaked at 19 litres each and, of the cows, 22 reached 32 litres and the remainder averaged 28 litres at peak yield. What is the approximate total amount of milk that the farmer could expect in a year?
17. A shepherd with 370 ewes says his true lambing percentage is 159. How many lambs were reared?
18. Which is the better result: (a) 107 lambs reared from 67 ewes served, or (b) 375 lambs reared from 251 ewes served?
19. If by early weaning at 3 weeks a sow can produce 3 litters of 9 piglets in 15 months, what is her production per year?
20. A litter of 8 pigs weighed 145 kg when they were weaned. Their average weight 140 days later was 91 kg. How much had each one gained on average per day? During the 140 days they had eaten 1·85 tonnes of meal. How much food did it take to put on 1 kg of liveweight gain?
21. A pen of intensive beef averaged 46 kg liveweight at birth and were sold at 12 months old weighing 432 kg. What was their average gain per head per day?
22. Calculate the deadweight of a baconer weighing 84 kg liveweight, a finished bullock weighing 550 kg liveweight and cull ewe of 60 kg liveweight.
23. One pig weighed alive 92·5 kg and its deadweight was 70 kg. Another weighed 68 kg alive and 47.5 kg deadweight. Which pig has the higher killing out percentage?
24. An intensive beef steer ate 1·35 tonnes of food from 12 weeks, when it was 100 kg liveweight, to 12 months, when it was slaughtered at 432

kg. How many kg of food were needed for each 1 kg of liveweight gain? If barley made up 70% of the ration, how much was used?

25. A pig from 18 kg to 54 kg liveweight ate 90 kg of a rearing ration and from 54 kg to 92 kg liveweight ate 144 kg of a finisher ration. What was the food conversion on each ration? What was the average food conversion throughout the whole period?
26. If the rearing ration in Question 25 costs 19·9p per kg and the finishing ration 17·5p per kg, what is the cost per kg liveweight gain for the (a) rearing period (b) finishing period (c) over the whole period from 18 kg to slaughter?
27. A beef animal weighing 530 kg and killing out at 56% made 208p per kg deadweight. How much did the farmer receive?
28. A farmer with a Friesian herd producing 62,000 litres in the month of April was informed that the average quality of the milk was 4.29% butterfat, 3·37% protein and 4·65% lactose, with a TBC in Band A. Using the values for each constituent quoted on page 51, calculate the gross total income for the month.
29. Compare the value of two pork pigs each weighing 64 kg liveweight, one of which was sold in the market for 94·5p per kg liveweight and the other to a wholesaler for 123·25p per kg deadweight.
30. How many Grade 2 baconers at the same liveweight must be sold to equal the returns from 50 Grade 1 pigs at the prices shown on page 59?

Chapter 4

Horticultural Calculations

Chapter 4

Horticultural Calculations

HORTICULTURE, like some aspects of agriculture, is an industry concerned with plants grown on or in areas of certain sizes, or in structures like glasshouses, cold frames or cloches which offer some protection and encouragement for the plants to grow out of season.

Many of the appropriate calculations will be similar to those for farm crops, although sometimes on a smaller scale, and include land and plot sizes, the size and volume of buildings, seed sowing and planting out dimensions, fertilisers and spraying.

AREA OF LAND

You should refer to page 18 to find out how the area of regular shapes is calculated. If you have a border of the shape shown in Fig. 1, it clearly consists of two rectangles, the areas of which can easily be worked out.

Similarly there will be other border shapes involving a curve (Fig. 2), and using the method for finding either the area of a circle

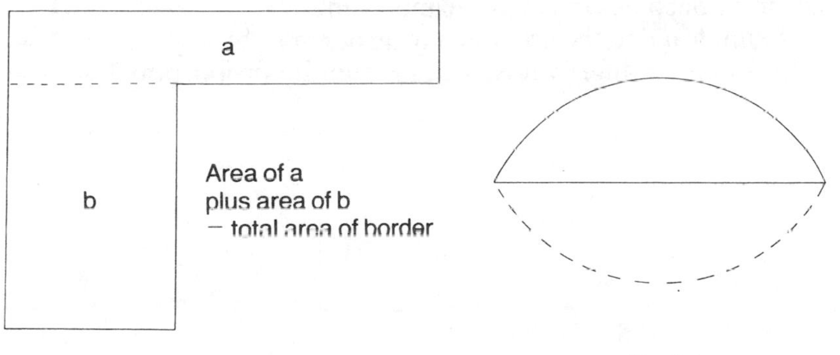

Fig. 1. *Fig. 2.*

or the area of an ellipse or an oval (see page 65), enables the area of a border with curved sides to be calculated.

Right Angles

There are, however, a few calculations of area which are rather special to horticulture. For instance, when cutting out a new border it may be necessary to mark out a right angle. If you take a length of string fixed to a peg at one end and with the other end you walk round in a complete circle you will have travelled through 360 degrees—written 360° (Fig. 3). If you walk only half way round then you will have travelled through 180° (Fig. 4) and you can see this represents the continuation of a straight line. If you walk only a quarter of the way round then this will be 90° (Fig. 5) and is called a right angle.

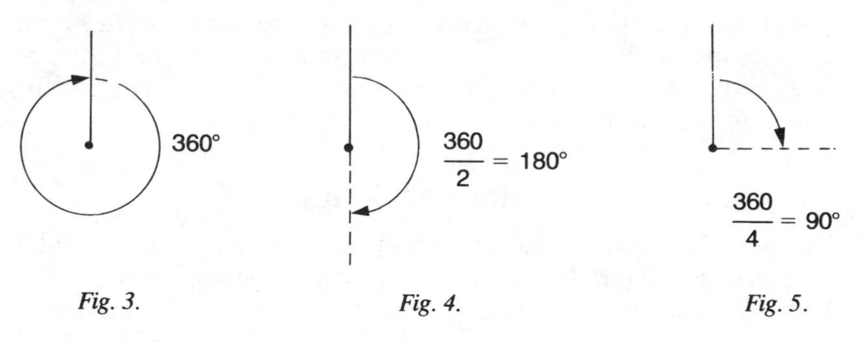

Fig. 3. *Fig. 4.* *Fig. 5.*

When setting out borders, you might judge the shape of a right angle roughly; with small borders a small error may not matter. But as the size of the border increases, any error will show up, and accuracy becomes much more important.

A simple aid to the accurate laying out of right angles is the 3, 4, 5 rule. In any triangle whose sides are in the proportion 3, 4, 5 the

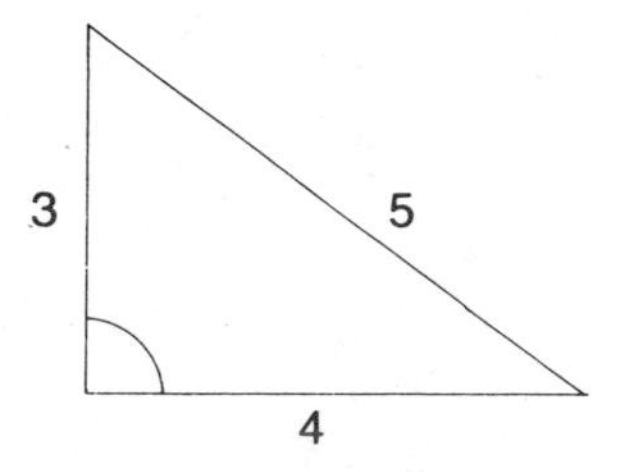

Fig. 6. Triangle.

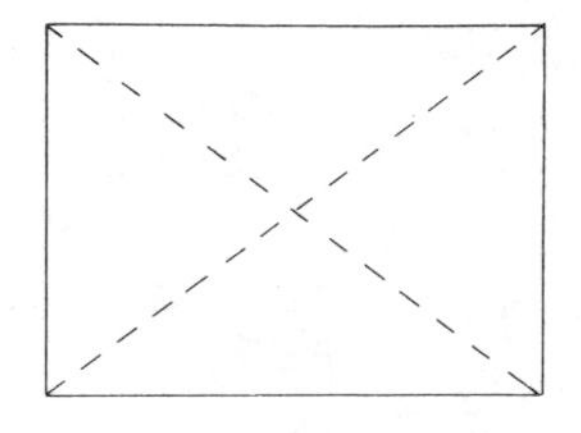

Fig. 7. Rectangle.

angle formed opposite the longest side will always be a right angle. It does not matter what units are used for the lengths—metres, centimetres, feet or yards. So long as the proportions are the same the rule holds good.

You may need to set out a larger border or bed than this. It does not matter what the measurements are, so long as they are in the same proportion. Provided you multiply your sides of length 3, 4, and 5 by the same number, you just increase the size of the triangle and still keep the right angle.

Example: You are setting out a rectangular border in a large garden, with two sides each measuring 12 metres, and two sides of 9 metres.

$$3 \times 3 = 9; \quad 3 \times 4 = 12; \quad 3 \times 5 = 15.$$

This gives our original 3, 4, and 5 each multiplied by 3. So with the sides of 12 metres and 9 metres, it needs a long side—in this case a diagonal—of 15 metres. Measuring the diagonals—the lines joining the opposite corners—provides a check on the accurate layout of any square or rectangular shape. The two diagonals should be equal in length.

Ellipses and Ovals

Other shapes often used in landscaping are the ellipse and the oval, both of which have curved sides but are not fully round like a circle.

With the ellipse, there is a centre point through which the longest lines of width and length both pass. At this point, both lines are cut exactly in half:

$$A = a \quad \text{and} \quad B = b.$$

In the oval only one of these lines—the longest line of width—is cut exactly in half by the other line, as can be seen in Fig. 9. All the same, the calculation of area for an oval is the same method as for an ellipse.

Remembering that the area of a circle is $\pi \times r \times r$ (πr^2), then the area of an ellipse or an oval is $\pi \times \frac{1}{2}$ longest length $\times \frac{1}{2}$ longest width. The areas of borders which are of a shape that is half an ellipse or an oval (such as Fig. 2. on page 63) can easily be calculated.

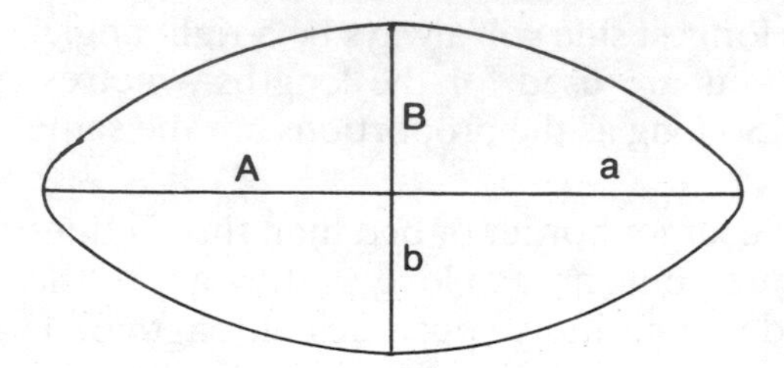

Fig. 8. Ellipse.

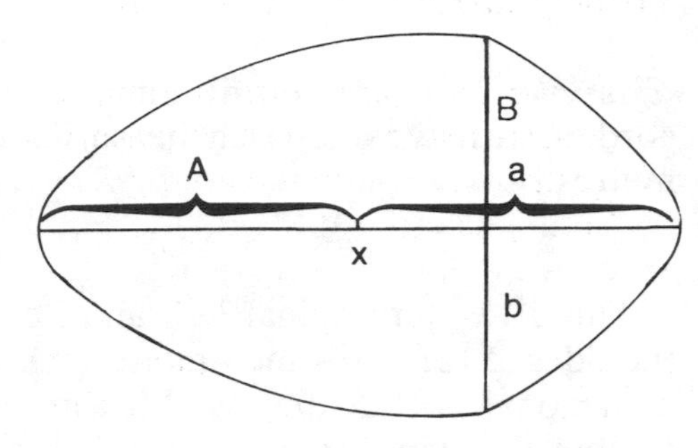

Fig. 9. Oval.

Example: A border in the shape of an ellipse has been cut out in a lawn so that the longer distance across is 4·5 m and the shorter width is 2·5 m. What is the area?

$$\frac{4{\cdot}5}{2} = 2{\cdot}25 \text{ m} \qquad \frac{2{\cdot}5}{2} = 1{\cdot}25 \text{ m}$$

The area will therefore be 3·14 (π) × 2·25 × 1·25 = 8·8 m^2, and the area of a border half the shape of this ellipse would be 4·4 m^2.

Perimeters

Where the intention is to plant up the edge of the border with edging plants it may be necessary to calculate the length of the outside edge or perimeter.

If the border has straight sides this is merely the sum of the separate sides which make up the shape. In a circular bed the perimeter is the circumference and is calculated from pi (π) × diameter (see page 19).

An ellipse or oval has a long and short diameter, and so the circumference is worked out as follows:

$$\frac{\text{Long diameter} + \text{Short diameter}}{2} \times \pi.$$

Another way in which the perimeter measurement can be used is for calculations of the boundary of a field or other area, for working out hedging or fencing. The method is the same whether used for horticulture or for farming purposes.

Example: A 1·5 hectare holding is rectangular in shape and its longest side is 150 m. It is to be fenced using stakes and wire. What is the distance round the holding?

Area is length × breadth, so 15,000 m^2 = 150 m × the shortest side.

$$\frac{15{,}000}{150} = 100 \text{ m, the length of the shortest side,}$$

$$2 \text{ sides at } 150 \text{ m} = 300 \text{ m},$$

$$2 \text{ sides at } 100 \text{ m} = 200 \text{ m}.$$

The total length round the holding—the perimeter—is 500 m.

Surface Area—Paving Slabs and Bench Pots

Another calculation frequently used is the estimation of surface area. Its use can best be understood in working out the number of concrete slabs required for paving an area associated with an amenity feature. Given the area to be paved and the size of each paving stone, it is fairly simple to find out how many will be needed.

Example: A rectangular area 9 m by 3 m is to be paved with slabs, each of which measures 0·6 m × 0·6 m. How many should be ordered?

Area to be paved $9 \times 3 = 27 \text{ m}^2$.

Size of one slab $0{\cdot}6 \times 0{\cdot}6 = 0{\cdot}36 \text{ m}^2$.

$$\frac{27}{0{\cdot}36} = 75 \text{ paving slabs to be ordered.}$$

A similar calculation is required to find the number of pots which can be placed on a bench. Both square and round pots can be placed in straight rows. On the other hand better use of available space will be made if the round pots are placed with alternate rows staggered.

Pots will occupy the space of their widest dimensions (usually across the top) and for a square pot this will simply be the square of one side. Round pots in straight rows will occupy an area equivalent to their diameter squared. If round pots are staggered the diameter is first squared and then multiplied by 0·9 to allow for the space the pots save as they fit against each other.

Example: How many 7·5 cm round pots will fit onto a bench 8 m^2 in (a) straight rows and (b) staggered rows?

The diameter of the pot is $7{\cdot}5 \div 100 = 0{\cdot}075$ m.
The area each will occupy in straight rows is:

$$0{\cdot}075 \times 0{\cdot}075 = 0{\cdot}0056 \text{ m}^2.$$

The total number of pots on the bench will therefore be:

$$\frac{8}{0{\cdot}0056} = 1{,}428 \text{ pots.}$$

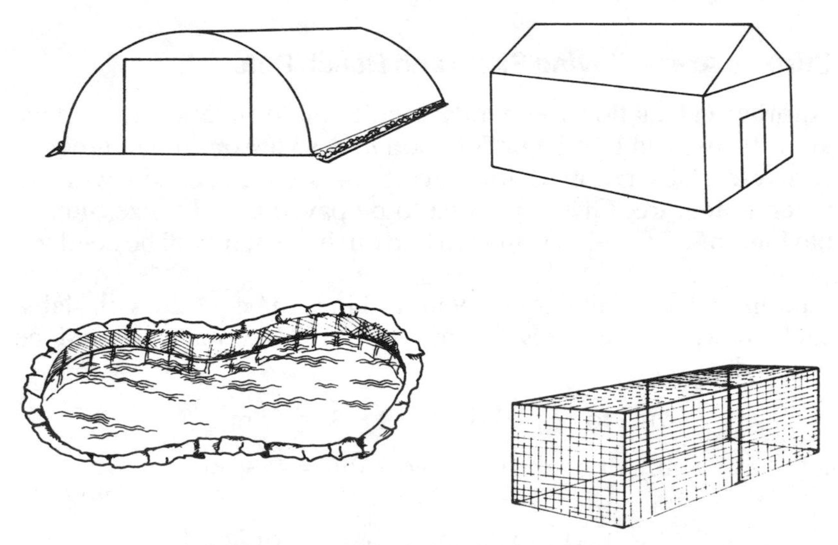

Examples where surface areas have to be calculated.

In staggered rows with space saved the number of pots increases:

$$0{\cdot}0056\ m^2 \times 0{\cdot}9 = 0{\cdot}00504\ m^2$$

$$\frac{8}{0{\cdot}00504} = 1{,}587 \text{ pots.}$$

Other examples where it is necessary to know the surface area would be:

- plastic tunnels (allow an overlap on each side)
- shading and interior insulation of glasshouses
- the area of a liner for a pond feature
- netting to make a fruit cage.

Example: What is the surface area of a fruit cage 8 m by 5 m, allowing for a working height inside of 2 m?

Two long sides	$8\ m \times 2\ m \times 2$	$= 32\ m^2$
Two short sides	$5\ m \times 2\ m \times 2$	$= 20\ m^2$
One top	$5\ m \times 8\ m$	$= 40\ m^2$.

Total surface area is $92\ m^2$.

If this netting can be purchased in rolls of $16\ m \times 2\ m$, how many rolls will be needed?

$$16\ m \times 2\ m = 32\ m^2.$$

$$\frac{92}{32} = 2{\cdot}9 \text{ rolls.}$$

Therefore three rolls of netting must be purchased.

VOLUME

Glasshouse Space

The most obvious use of volume is in calculating the interior space of glasshouses, particularly for working out how much fumigant is required to achieve thorough fumigation. Glasshouses can be

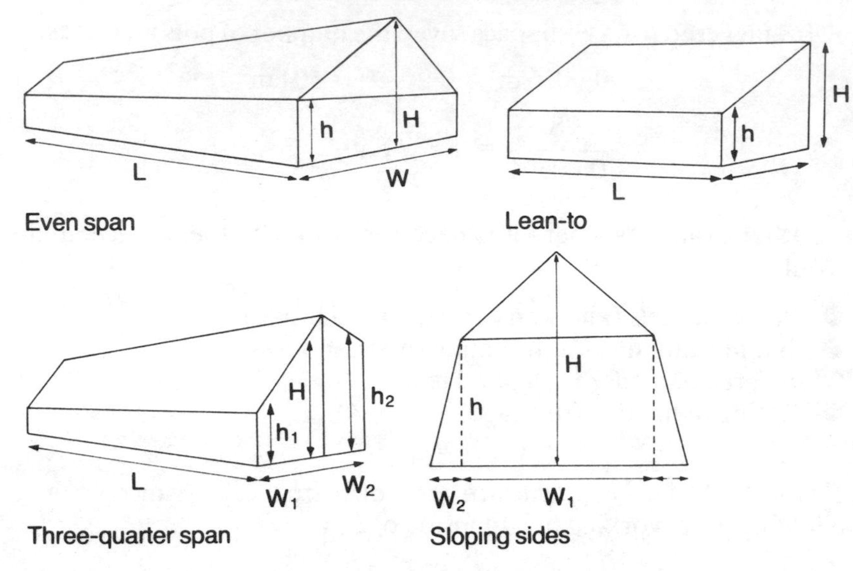

Glasshouse shapes.

designed in a number of different ways to suit the circumstances. They can be free-standing, joined together in series or built as a lean-to against a wall. Above are four examples of glasshouse design.

The volume for the even span and lean-to can be worked out in the same way. They have both got length (L), width (W) and two different heights, one to the eaves (h) and one to the ridge (H).

A short cut that you can use is to calculate the average height, which will be:

$$\frac{h + H}{2}.$$

This average height can then be multiplied by length and width to find the volume. The whole calculation is shown as:

$$\frac{h + H}{2} \times L \times W = \text{cubic capacity}.$$

For a three-quarter span glasshouse, the volumes of two halves must be worked out separately and then added together:

$$\frac{h_1 + H}{2} \times L \times W_1, \qquad \text{plus } \frac{h_2 + H}{2} \times L \times W_2.$$

This will give the total cubic capacity of the glasshouse.

Many glasshouses have sloping sides and an end profile is illustrated. In this case there are two extra equal-sized triangles which must be added in to the surface area of the end of the glasshouse.

The area of a triangle is:

$$\frac{B \times H}{2}$$

(see page 18), but because these are two identical triangles you need only calculate B × H. The total surface area of the end of a sloping-sided glasshouse then becomes:

$$\frac{h + H}{2} \times W_1, \quad \text{plus} \quad h \times W_2.$$

Example: An even span commercial glasshouse measures 40 m long and 17 m wide. The height to the eaves is 3·4 m and to the ridge it is 4·6 m. What is the cubic capacity?

$$\text{Average height} = \frac{4{\cdot}6 \text{ m} + 3{\cdot}4 \text{ m}}{2} = 4 \text{ m}$$

$$4 \text{ m} \times 17 \text{ m} \times 40 \text{ m} = 2{,}720 \text{ m}^3.$$

If a local supplier sells fumigants each of which is effective in 280 m^3 of space, how many will be needed to fumigate the whole glasshouse?

$$\frac{2{,}720}{280} = 10 \text{ fumigants.}$$

The same principle applies to stores for produce, e.g. fruit cold stores. The cubic capacity represents storage space and gives you the opportunity to estimate how much produce is left in store at any time.

Cubic capacity is equally important in the laying of concrete paths and standings. Over hard core, most paths would need to be 0·1 m deep to carry normal weights, and for heavy equipment up to 0·15 m deep.

If you had to build a path round the perimeter of a new building you would need to remember that it would extend beyond the building and the extra area at each of the four corners must be included.

Example: How much ready-mix concrete should be ordered to lay a path 2 m wide round the perimeter of a new store which itself measures 8 m by 6 m?

The two shorter sections are each 6 m × 2 m in area. On the larger areas allow 2 m extra each end for the width of the path, making two sections 12 m × 2 m. The volume of concrete required is:

$$2 \times (6\text{ m} \times 2\text{ m} \times 0{\cdot}1\text{ m}) + 2 \times (12\text{ m} \times 2\text{ m} \times 0{\cdot}1\text{ m})$$

$$2{\cdot}40\text{ m}^3 + 4{\cdot}80\text{ m}^3 = 7{\cdot}20\text{ m}^3.$$

TEMPERATURE

Conversions

There may be a little confusion at the present time, as both Centigrade (C) and Fahrenheit (F) scales are used. Make sure which one is being used, and that it is clearly stated. Centigrade is also called Celsius.

It is often useful to know how to convert from one scale to the other.

To convert °C to °F: $(°C \times 1{\cdot}8) + 32 = °F.$

To convert °F to °C: $(°F - 32) \div 1{\cdot}8 = °C.$

Some useful markers are:

body temperature	37°C
a warm day	20°C
inside greenhouses	20°C
a cool day	5°C
normal frost	0°C to −4°C.

Two very useful relationships are:

10°C = 50°F

16°C = 61°F (note the reversed numbers).

Boiler Outputs

In order to confirm the correct size of boiler for use with a glasshouse it may be necessary to work out the required heat output of the boiler which is measured in watts (W) and kilowatts (kW)— 1,000 watts. The requirement depends on the surface area of the glass, the temperature outside the glasshouse and the temperature needed inside.

(a) First calculate the total surface area of glass in m^2 (see page 71).
(b) Work out the temperature lift which is the difference between the average lowest expected outside temperature in winter and the temperature required inside.
e.g. If temperatures outside can fall on average to −10°C and the temperature needed inside is 15°C then the lift is 25°C.
(c) Make an allowance for loss of heat through the different surfaces, usually a constant of 8 watts per square metre per °C temperature lift.*
(d) Finally surface area of glass in $m^2 \times$ temperature lift in °C × 8 for heat loss = heat output of the boiler in watts.

Example: What will be the heat output of a boiler required for a glasshouse with the following measurements: 8 m long; 3 m wide; 2 m height to eaves; 3 m height to ridge; 1·8 m slant height of roof. Assume the temperature lift to be 25°C.

Surface area is made up of:

2 ends $3 \times \frac{(2+3)}{2} \times 2 = 15\ m^2$

2 sides $2 \times 8 \times 2 = 32\ m^2$

2 roofs $1{\cdot}8 \times 8 \times 2 = 29\ m^2$.

Total surface area is $76\ m^2$.

Heat output of boiler will be $76 \times 25 \times 8 = 15{,}200$ watts.

Normally, it is recommended that ¼ to ⅓ of this figure should be added, to deal with emergencies and to avoid running the boiler flat out in cold weather.

*You should be aware that heat loss in a glasshouse varies; through glass it is 5·7 watts; through brick it is 3·4 watts and through cracks and doors it is 1·4 watts. An arbitrary figure of 8 watts per m^2 per °C temperature lift has been used to take account of the relative contributory losses through different surfaces of a standard glasshouse.

QUANTITY CALCULATIONS

Fertilisers

The principle of making the comparison in cost between two or more fertilisers containing different amounts of the same plant food is exactly the same as the explanation given on page 34.

The method of selecting the correct compound fertiliser for the special requirements of each crop also follows the same procedure as outlined in the chapter on farm crop calculations (see page 35). As a reminder, here is a step-by-step summary.

(a) Find out the needs of the crop in terms of kg of N, P and K.
(b) Convert these requirements to a plant food ratio.
(c) From the range of fertilisers available, choose one which has a plant food ratio similar to the crop needs.
(d) Calculate how much to apply per hectare, or part hectare, to give the right amounts of plant foods or as near as possible.

On smallholdings the areas to be covered and consequently the amounts applied are often relatively small. Useful conversions to remember which are reasonably accurate are:

$$35 \text{ g per m}^2 = 1 \text{ oz per square yard}$$

$$0{\cdot}5 \text{ kg per m}^2 = 1 \text{ lb per square yard.}$$

Example: A flower border measuring 2 m by 3 m requires 70 g per m^2 of nitrogen, and the fertiliser available is 21% sulphate of ammonia. How much fertiliser will be applied to the border?

$$6 \text{ m}^2 \times \frac{70 \text{ g}}{1{,}000} \times \frac{100}{21} = 2 \text{ kg of sulphate of ammonia.}$$

If a 34·5% nitrogen fertiliser was available instead, how much would be needed?

$$6 \times \frac{70 \text{ g}}{1{,}000} \times \frac{100}{34{\cdot}5} = 1{\cdot}22 \text{ kg of } 34{\cdot}5\% \text{ N fertiliser.}$$

Liming

The principles of liming follow those referred to on page 38. The various types of lime mentioned are similarly available to the horticulturalist and the neutralising values remain the same.

Quick-lime or burnt lime is the most concentrated and less is required to achieve the same improvement in soil lime status, but it is unpleasant to handle. A popular form of liming material is slaked lime and this is applied at 270 g per m^2 in normal circumstances of soil acidity.

Example: An area of the vegetable garden measuring 18 m by 4·8 m shows marked signs of acidity. How much slaked lime should be purchased to correct the problem?

$$\frac{18 \times 4{\cdot}80 \times 270\text{ g}}{1{,}000} = 23{\cdot}3\text{kg of slaked lime.}$$

Remembering that the neutralising values of slaked lime and ground limestone are 75 N.V. and 50 N.V. respectively, how much ground limestone would be needed to have the same neutralising effect, and at what rate would it be spread per m^2?

$$23{\cdot}3\text{ kg} \times \frac{75}{50} = 35\text{ kg of ground limestone to be purchased.}$$

$$270\text{ g} \times \frac{75}{50} = 405\text{ g of ground limestone spread per m}^2.$$

Seed Rates and Planting Distances

Because of the vast range of plants grown by horticulturalists including vegetables, flowers, shrubs and fruit, it is not practical to quote seed rates and planting distances. These can be easily obtained from suitable text books. It is important to be able to apply the information and make the correct calculation of available space and plants required for any given area of land.

Just as in agriculture, there is a correct density of planting so that each plant obtains sufficient moisture, nutrients and light without overcrowding or being so thinly spaced as to leave the land bare and encourage the growth of weeds. For some plants the seed is sown thinly along a row and after emergence the young seedlings may or may not be thinned out. With others, the seed or other plant materials are spaced further apart at distances where the plant will remain until maturity. Other plants will be transplanted as seedlings at the correct distances apart.

If there are clearly defined planting measurements, it is then possible to work out the number of rows or plants which can be established in a given area.

Example: In a border 8 m × 2·5 m, roses are to be planted on the square at 0·6 m. How many plants should be ordered?

Each rose will take up a space 0·6 m × 0·6 m = 0·36 m²
The area of the border is 8 m × 2·5 m = 20 m²

Therefore the number of roses required to complete the border is:

$$\frac{20}{0{\cdot}36} = 55 \text{ roses.}$$

This method can be applied to any area for any plants with a recognised planting distance.

Example: If new raspberry canes are planted 0·3 m apart and are sold in bundles of 50, how many will be needed to plant 3 rows across a plot each row of which measures 9 m?

$$\frac{3 \times 9}{0{\cdot}3} = 90 \text{ canes}$$

$$\frac{90}{50} = 2 \text{ bundles to be ordered.}$$

Grass seeds mixtures used for sowing lawns present a certain calculation problem. The seed from the different grasses vary in size. Thus for any given weight—whether a gram or a kilo—the number of seeds varies from one species (or variety) of grass to another. Thus, the correct balance of a grass seeds mixture must take into account the number of seeds per gram for each type of grass in the mixture.

Grass	*Approximate no. seeds per gram*	*Ratio*	*Multiply by*	*Result*	*% by weight*
Lolium perenne	500	20	3	60	55
Poa pratensis	4,000	2·5	3	7·5	10
Festuca rubra	1,000	10	3	30	30
Agrostis tenuis	10,000	1	3	3	5

The procedure for formulating a seeds mixture is as follows:

(a) For each grass in the mixture, divide the number of seeds per gram into the largest figure in the same column. Thus, *Lolium perenne* (500) divides into *Agrostis tenuis* (10,000) 20 times. Thus the figure 20 is put into its place in the column headed 'ratio'.
(b) Find the total of all the numbers in the 'ratio' column, in this case 33·5, and divide this total into 100. This time, it gives an answer of 3.
(c) Multiply each ratio by this figure of 3, to give a percentage by weight for each of the grasses in the mixture. These percentages then appear in the column headed 'result'.
(d) Normally, numbers are adjusted, if necessary, to give a set of round figures which add up to 100 (in the final column).

Carrying out this procedure ensures that each grass contributes a similar number of seeds to the mixture, in spite of the variations in seed size and weight.

The same idea is applied in grass and clover mixtures for farm use. Small quantities of Timothy and White Clover are included, for example, in a mixture with the Ryegrasses. Thus, although the first two are very small seeds, there is no shortage of these two species when the plants in the mixture start to grow.

Spraying and Spray Chemicals

Because sprays are costly and can be harmful to plants and beneficial insects if used to excess, it is important to make sure that the sprayer is working correctly and applying the correct amount. This calls for calibration and here is a method for calibrating the knapsack sprayer. It helps to remember that:

1 kilogram = 1,000 grams

1 litre = 1,000 millilitres

1 hectare = 10,000 square metres. (See pages 106 and 107.)

(a) Put a litre of water in the sprayer.
(b) Spray it out at the correct spraying speed and boom height.
(c) Once empty, measure the area covered by a litre of water.
(d) Multiply this area by the capacity of the knapsack sprayer when full (usually 20 litres).

Knapsack sprayers are often sold with a range of different nozzle sizes, usually coloured. A separate calibration must be carried out for each nozzle in order to ensure that the application is accurate.

Example: If 10 m^2 are covered by 1 litre of water then 200 m^2 will be covered by a full container i.e. $10 \times 20 = 200\ m^2$.

The next operation is to measure out accurately the amount of chemical for a full tank.

Example: If the application rate of a spray chemical is 11 litres per ha then the rate per m^2 will be:

$$\frac{11{,}000\ \text{ml}}{10{,}000\ \text{m}^2} = 1{\cdot}1\ \text{ml per m}^2.$$

Since the sprayer when full will spray 200 m^2, the amount to be added to 20 litres of water in the tank is:

$$1{\cdot}1 \times 200 = 220\ \text{ml of spray chemical.}$$

As for comparisons of cost between two or more sprays which claim to do the same job, these can be made in exactly the same way as described on page 39.

Liquid Feeding

The application of plant foods in liquid form is becoming a fairly common practice in modern glasshouse production. The aim for any crop is to apply liquid feeds which will have the important plant nutrients combined in the correct ratio and at the correct strength.

These liquid feeds are obtained from stock solutions which are prepared by adding a solid fertiliser to water. Here is a list of fertilisers with their plant food content that are frequently used as a basis for liquid feeding.

Ammonium nitrate	35·0% N	
Sulphate of ammonia	21·0% N	
Urea	46·0% N	
Potassium nitrate	13·8% N	46·5% K_2O
Sodium nitrate	16·0% N	
Mono-ammonium phosphate	11·0% N	48% P_2O_5.

The concentration of a stock solution is expressed as a percentage of the weight of fertiliser in a volume of water (W/V).

Example:

200 g in 1 litre (1,000 g) = 20% W/V

1 lb in 1 gallon (10 lb) = 10% W/V.

Clearly if one litre of stock solution is insufficient and, for example, 5 litres are required, then 1,000 g or 1 kg in 5 litres is still 20% W/V since both numbers have been multiplied by 5.

From a stock solution, the grower will carry out dilutions by adding more water to it at the correct amounts. A typical rate of dilution used is 1 part of stock solution to 200 parts of water. The final strength of the dilution is referred to as parts per million (ppm) of the plant nutrient.

With this basic information, it is now possible to relate the concentration of the stock solution to ppm of the plant nutrient in the final dilution.

ppm of the nutrient =

$$\frac{\text{fertiliser weight in g} \times \%\ \text{nutrient in solid} \times 10^{*}}{\text{rate of dilution}}$$

Example: 150 g of ammonium nitrate in a stock solution was diluted by 250 parts of water. What were the ppm in the dilution?

The plant food content of ammonium nitrate is 35%, therefore:

$$\frac{150\ (\text{g}) \times 35\ (\%) \times 10}{250\ (\text{dilution})} = 210\ \text{ppm}.$$

Conversely, the weight of fertiliser to be added to a known quantity of water, usually 1 litre or 1 gallon, to make up a stock solution can be worked out if the ppm of the plant nutrient and the extent to which it has been diluted are both known.

g of fertiliser per litre =

$$\frac{\text{ppm of nutrient diluted} \times \text{dilution}}{\%\ \text{nutrient in solid} \times 10^{*}}$$

*If the fertiliser is weighed out in lb then change the asterisked number to 1,000, or if it is weighed in oz change it to 62·5.

Example: What quantity of sulphate of ammonia must be added to 1 litre of water to make up a stock solution if the plant nutrient is at 250 ppm and the dilution was made at 1 in 200?

$$\frac{250\ (\text{ppm}) \times 200\ (\text{dilution})}{21\ (\%) \times 10^*} = 238 \text{ g in one litre for the stock solution.}$$

By combining the two formulae, it is possible to work out the preparation of a stock solution from a combination of two solid fertilisers which contain different plant foods.

Example: Calculate the amounts of potassium nitrate and ammonium nitrate to be added to 5 litres of water as a stock solution which when diluted at 1 part in 200 provides 100 ppm of N and 75 ppm of K_2O.

$$\text{For the } K_2O \text{ as g per litre } \frac{75\ (\text{ppm}) \times 200\ (\text{dilution})}{46{\cdot}5\ (\%) \times 10} = 32{\cdot}3 \text{ g.}$$

The N supplied in this amount would be:

$$\frac{32{\cdot}3\ (\text{g}) \times 13{\cdot}8\ (\%) \times 10}{200\ (\text{dilution})} = 22{\cdot}3 \text{ ppm N.}$$

The additional N required will be $100 - 22{\cdot}3 = 77{\cdot}7$ ppm N, therefore:

$$\frac{77{\cdot}7\ (\text{ppm}) \times 200\ (\text{dilution})}{35\ (\%) \times 10} = 44{\cdot}4 \text{ g of ammonium nitrate.}$$

To make up 1 litre of stock solution it would be necessary to add 32·3 g of potassium nitrate and 44·4 g of ammonium nitrate. For 5 litres, the amounts added would therefore be 161 g and 222 g respectively.

HORTICULTURAL PROBLEMS

1. A circular flower bed of 3·65 m diameter has been cut out of an existing lawn and the grass surface is to be used as turfs for a path elsewhere. Ignoring wastage, what is the total area of turf available?

*If the answer is to be given in lb or oz, follow the rule given on page 79.

2. In a rectangular-shaped holding of 2·16 hectares, two sides are 50% longer than the opposite pair. What are the outside dimensions of the holding?
3. What is the area of a triangle whose sides are 3, 4 and 5 metres? If the sides are doubled in each case, by how much does the area of the triangle increase in size?
4. How many 10 cm round pots will fit on a 2 m^2 bench if they are: (a) in straight rows, (b) staggered? How many 5 cm square pots will fit on the same bench?
5. A beech hedge is to be planted round one long side and two short sides of a garden which measures 130 m by 70 m. If the distance apart of the hedging material is 0·45 metre, how many plants will be needed to complete the enclosure? If the purchase price is £6·60 per 10 plants, what is the cost of the order?
6. An elliptical bed has been created as a garden feature. At its widest it is 3·6 m across, and the maximum length is 25% more than this. What will be the area of the bed?
7. A neighbour is taking delivery of a load of ready-mix concrete and estimates he will not need 2·5 m^3 which he offers to you for a patio you are planning. If it is laid at 0·1 m deep, what would be the size of a square patio which you would be able to lay?
8. A circular pond of 3 m diameter and dug vertically down to a depth of 60 cm requires a plastic lining so that there is at least 15 cm overlap on to the surrounding soil surface. What is the minimum size of lining which must be obtained?
9. A glasshouse measures 30 m long, 6·5 m wide, 2·2 m to the eaves, 3·7 m to the ridge and 3·6 m slant height of roof. Bubble plastic insulation is available in rolls 1·5 m wide and 50 m long at a cost of £42 per roll. If the glasshouse is to be insulated all the way round up to the eaves, how many rolls must be purchased and what is the total cost?
10. For the size of glasshouse in the previous question, what must be the heat output of the boiler if the coldest temperature in winter is likely to be −12°C and the ideal operating temperature is 18°C.
11. It is predicted that the daily temperature will reach 82°F in the shade. What is this in Centigrade (degrees Celsius)?
12. If normal plant growth begins at 7°C, what will this be in Fahrenheit to the nearest whole number?
13. An expanse of Venlo-type glass is made up of a number of identical units each with its own ridge but without the internal sides. The dimensions of one unit are 3·2 m wide × 36·6 m long × 2·3 m to the eaves × 3·1 m to the ridge and there are 8 of these joined together.

Calculate the total volume enclosed within and the number of fumigants required each of which is able to fumigate 280 m^3 of space.

14. A plot of 2·2 hectares has been divided up to grow equal areas of Pick Your Own/Farm Shop vegetable crops, each of which has its own requirements for plant foods in kg per ha. The crops are potatoes (185:185:260); peas for picking (0:50:50); Brussels sprouts (160:80:240); autumn-sown bulb onions (70:180:180) with 140 kg of N as a top dressing in the spring. Using information given in Chapter 2, choose the correct fertiliser for each crop and the amounts which should be used per hectare.
15. For the plot referred to in the previous question, lime is applied each year to the area reserved for the peas. How much ground limestone would be needed to give the land an adequate dressing? What would be the equivalent amounts of (a) burnt lime (b) slaked lime if they were used instead?
16. A border is to be planted up with daffodils to cover 75% of the space and polyanthus in the remaining area. If the daffodils are planted 15 cm × 15 cm apart and 400 bulbs were purchased, how many polyanthus at 30 cm × 30 cm apart are required to fill the area reserved for them?
17. If the total area of the border in the previous question is to be planted up with dahlias for the second half of the growing season, how many tubers are needed if they are normally planted at 45 cm × 45 cm?
18. A rectangular area of soft fruit is being planted so that in order across the plot there are: 3 rows of raspberries at 1·2 metres apart; 3 rows each of 6 blackcurrant bushes at 1·5 metres by 1·5 metres; and 4 rows of strawberries at 0·8 metres between the rows. Outside the rows of fruit (around the plot) is a distance of 1 metre at each side and 1 metre at each end. Between the blackcurrants and the other fruit on each side is 1·5 metres. How much land will this soft fruit planting occupy?
19. If this area of soft fruit (in question 18) is to be netted to a working height of 2 metres and over the top, what will be the total surface area of net?
20. If top fruit is normally spaced at 4·5 m × 4·5 m, what size of plot will be necessary to plant 40 bush apple trees?
21. A ryegrass-free fine turf mixture is made up of *Poa pratensis*, *Festuca rubra* and *Agrostis tenuis*. What percentage by weight of the three constituents will appear in the final mixture?
22. Calibration of a normal-sized knapsack sprayer showed that 1 litre of water covered 15 m^2 of land. A spray chemical is to be used at the rate of 7 litres per hectare. How much must be added to a full tank?

23. A stock solution containing mono ammonium phosphate was diluted to 1 part in 250 and provided 80 ppm of N. (a) What quantity in grammes of fertiliser was added to 1 litre of water to make up the stock solution? (b) How many ppm of P_2O_5 is present in the dilution?

Chapter 5

Machinery Calculations

Chapter 5

Machinery Calculations

MOST OF these calculations are capable of being done with a good degree of accuracy, thus precise measurements of size, weight, speed, temperature, etc. are needed.

CALIBRATION

Calibration means checking the application rates of drills, sprayers and fertiliser distributors. By using standard formulae, you can calculate the application rate of a machine while it is running either in the yard or in a building.

Corn Drills

Many modern drills have a crank handle for checking the application rate. A check in the instruction book will tell you how many times to turn the handle to cover the equivalent of a specified area. For other drills, it is necessary to jack the drill driving wheels clear of the ground, making them safe with axle stands, and then turn the driving wheel(s) a certain number of times to cover a chosen area. A simple formula is used to calculate the number of times the wheel must be turned to cover the equivalent of $\frac{1}{25}$ hectare or $\frac{1}{10}$ acre.

First it is necessary to calculate the sowing width of the drill. This is found by multiplying the row width by the number of coulters.

Example:
The drill has 21 rows at 16 cm spacing so the sowing width is:

$$\frac{16 \times 21}{100} \text{ metres} = 3{\cdot}36 \text{ metres}$$

or

the drill has 21 rows at 7 inch spacings so the sowing width is:

$$\frac{7 \times 21}{12} \text{ feet} = 12{\cdot}25 \text{ feet.}$$

Next, measure the circumference of the driving wheel in feet or metres, using the formula: Circumference = Diameter × $\frac{22}{7}$. The Circumference of a wheel with a Diameter of 1·2 metres will be:

$$\frac{1{\cdot}2 \times 22}{7} = 3{\cdot}77 \text{ metres.}$$

Now that the sowing width and wheel circumference have been found, use one of these formulae to find the number of turns required to cover the stated area.

The number of turns for:

$$\frac{1}{25}\text{ hectare} = \frac{400}{\underset{(m)}{\text{Sowing width (SW)}} \times \underset{(m)}{\text{wheel circumference (C)}}}$$

or

the number of turns for:

$$\frac{1}{10}\text{ acre} = \frac{484 \times 9}{\underset{(ft)}{\text{Sowing width (SW)}} \times \underset{(ft)}{\text{wheel circumference (C)}}}$$

Example 1

Find the number of turns of the drill wheel to cover $\frac{1}{25}$ hectare if the sowing width is 4 metres and the wheel has a circumference of 3·1 metres.

Number of turns for:

$$\frac{1}{25}\text{ hectare} = \frac{400}{\text{SW }(m) \times \text{C }(m)} = \frac{400}{4 \times 3{\cdot}1}$$

$$= 32{\cdot}25 \text{ turns.}$$

Example 2

Find the number of turns of the drill wheel to cover $\frac{1}{10}$ acre if the sowing width is 10·5 feet and the wheel circumference is 12·5 feet.

Number of turns for:

$$\tfrac{1}{10}\text{ acre} = \frac{484 \times 9}{\text{SW}\,(ft) \times \text{C}\,(ft)} = \frac{484 \times 9}{10{\cdot}5 \times 12{\cdot}5}$$

$$= 33 \text{ turns.}$$

To carry out the calibration, turn the driving wheel the calculated number of times. Make sure the drill is in gear and at the required setting. Collect the seed from the coulters on sacks or other material and, when the turning is completed, weigh the seed delivered by the drill and multiply this by 25 to obtain the seed rate per hectare, or by 10 to find the seed rate per acre. If the seed rate is found to be inaccurate, alter the setting slightly and repeat the procedure until the correct setting is found. Make a permanent record of the drill output at the various settings required.

Crop Sprayers

A static application rate check can be carried out for crop sprayers. The same calibration check can also be used for fertiliser broadcasters. The application rate is checked by running the sprayer for a calculated period which is equivalent to running across a field at a chosen speed with a measured spraying width.

To find the running time to spray a chosen area use the following formula:

$$\underset{(\text{minutes})}{\text{The time to spray 1 hectare}} = \frac{600}{\underset{(m)}{\text{Spraying width (SW)}} \times \underset{(km/hr)}{\text{Speed (S)}}}$$

The spraying width is found by multiplying the space between the jets by the number of jets. Remember to convert the figure to metres.

or

$$\underset{(\text{minutes})}{\text{The time to spray 1 acre}} = \frac{165}{\underset{(yd)}{\text{Spraying width (SW)}} \times \underset{(mph)}{\text{Speed (S)}}}$$

Example 1
Find the time required to run a sprayer to calibrate the machine for one hectare if the spraying width is 20 metres. Forward speed to be 6 km/hr.

$$\text{Time required for 1 hectare (minutes)} = \frac{600}{SW\,(m) \times S\,(km/hr)}$$

$$= \frac{600}{20 \times 6} \text{ minutes}$$

$$= 5 \text{ minutes.}$$

Example 2
Find the time required to run a sprayer to calibrate the machine for one acre if the forward speed is 5 mph and it has 24 nozzles at 20-inch spacing.

$$\text{Spraying width} = \text{number of nozzles} \times \text{nozzle spacing}$$

$$= 24 \times 20 \text{ inches}$$

$$= 480 \text{ inches} = 13{\cdot}33 \text{ yards.}$$

$$\text{Time required for 1 acre (minutes)} = \frac{165}{SW\,(yd) \times S\,(mph)}$$

$$= \frac{165}{13{\cdot}33 \times 5} = 2{\cdot}48 \text{ minutes.}$$

Having calculated the required running time to cover one hectare or one acre, fill the sprayer tank to a known mark—or full to the brim—run the machine for the calculated period and then re-fill the tank. By measuring the quantity of water needed to re-fill to the mark, the exact quantity of liquid sprayed can be found.

Fertiliser Spreaders

Exactly the same procedure can be used for calibrating a fertiliser broadcaster. Find from the instruction book the effective width of spread. Use this measurement instead of the spraying width in the formula to calculate the running time needed to cover one hectare or one acre.

Checking Application Rates in the Field

This can be done by measuring an area of the field, e.g. one hectare or one acre. Put the required amount of grain, fertiliser or chemical into the hopper or tank and drive until the area is completed. If the setting is correct the material will be used up at the same time as the area is finished.

Using the following method it is simple to mark out a convenient plot with an area of 1 hectare (10,000 square metres). Choose a width which is a multiple of the sowing or spraying width. For example, a sprayer with a working width of 30 metres could have a plot area for the test of four bouts wide giving 120 metres. To find the length of the test plot:

$$\text{Length of plot} = \frac{\text{Area of plot}}{\text{Width of plot}}$$

$$= \frac{10{,}000}{120} \text{ metres}$$

$$= 83{\cdot}3 \text{ metres.}$$

For an acre plot (4,840 sq. yds) with a working width of 12 yards, choose, for example, a plot width of 48 yards. The length of the test plot will be:

$$\frac{4{,}840}{48} = 100{\cdot}8 \text{ yards.}$$

After calculating the dimensions for either an acre or a hectare plot it is then a simple matter to mark out the area, put enough seed, fertiliser or water in the hopper or tank and then work the area. On completion of the acre or hectare, you can check that the right amount of material has been used.

Checking Forward Speed

With all machines driven by the tractor power take-off shaft which apply materials such as spray chemicals to the soil, the forward speed must be set according to the maker's instructions. It is possible to check the forward speed of a tractor by measuring the time taken to travel a fixed distance. This is also a means of checking the accuracy of the tractor meter which reads off the forward speed.

Use a tape measure to mark out 100 metres if checking the speed in kilometres per hour or 100 yards if checking a speed in miles per hour. Most tractor meters give the forward speed in miles per hour. Mark the ends of the measured distance with a stake.

Select the forward gear which, according to the tractor instruction book, gives the required speed. Set the throttle to give a power take-off speed of 540 rpm or some other speed which may be required for correct implement operation. You should start the test well before reaching the mark to ensure that the tractor is running at the correct speed while travelling the measured distance. Take a careful timing of the period taken to drive between the stakes and then refer to the charts below to check the tractor speed.

Checking speed over 100 metres

speed (km/hr)	3	4	5	6	7	8	9	10	11	12
time (seconds)	120	90	72	60	51	45	39	36	33	30

Checking speed over 100 yards

speed (mph)	2	$2\frac{1}{2}$	3	$3\frac{1}{2}$	4	$4\frac{1}{2}$	5	6	7	8
time (seconds)	102	82	68	58	51	45	41	34	29	26

Rates of Work

The amount of work done by a tractor and implement will vary considerably depending on working width, forward speed and time taken up for turning at headlands. Farmers are using both hectare and acre units of area but in most cases forward speed is measured in mph. Using the following formula, you can calculate the rates of work by any machine in either acres per hour or hectares per hour.

$$\underset{(\textit{acres per hour})}{\text{Rate of work}} = \frac{\text{working width } (\textit{ft}) \times \text{speed } (\textit{mph})}{10}$$

This formula allows 20% extra time for stoppages and turning at the headlands.

Example:
Find the output in acres per hour when using a corn drill with a working width of 12 feet with the tractor travelling at 5 miles per hour.

$$\text{Work rate} = \frac{\text{working width} \times \text{speed}}{10}$$

$$= \frac{12 \times 5}{10}$$

$$= 6 \text{ acres per hour.}$$

To convert the work rate to hectares per hour divide the rate in acres per hour by 2·47.

$$\text{Work rate} = \frac{6}{2{\cdot}47} = 2{\cdot}43 \text{ hectares per hour.}$$

Wheel Slip

Whenever a tractor is at work there will be some wheel slip. It is possible to find out how much slip occurs by taking a few simple measurements followed by a calculation.

First, measure the circumference of the tractor rear wheel. With the tractor standing on concrete, make a mark at the bottom of the tyre wall and continue the mark on to the concrete. Drive forward one revolution of the rear wheel and, using the mark on the tyre as a guide, make another mark on the concrete. Measure the distance between the marks to find the actual distance travelled (the circumference) by the rear wheel.

The tractor can now be taken to a field with an implement which is put into work. Using the mark on the tyre and a couple of canes, you can measure the distance travelled per wheel revolution when the tractor is under load. To get a more accurate result, allow the tractor wheel to turn ten revolutions between the first cane and the second cane.

Now, wheel slip can be calculated by measuring the distance travelled in the field and comparing it with the distance travelled on the concrete with no load.

Example:
A tractor with no load when tested on concrete travels a distance of 69 metres in ten revolutions of the rear wheel. A field test when ploughing gives a distance of 58·65 metres travelled in ten revolutions. Find the percentage wheel slip.

Difference in distance travelled

$$\text{in 10 revolutions} = 69 - 58{\cdot}65 \text{ metres}$$

$$= 10{\cdot}35 \text{ metres.}$$

$$\% \text{ wheel slip} = \frac{\text{distance lost to wheel slip}}{\text{distance travelled with no slip}} \times 100$$

$$= \frac{10{\cdot}35}{69} \times 100$$

$$= 15\% \text{ wheel slip.}$$

This calculation can also be done by using yards as the unit of measurement instead of metres.

Gear and Pulley Speeds

There is a relationship between the speed and diameter of a pair of belt pulleys. There is also a relationship between the speed and number of teeth on a pair of gears or a pair of sprockets used as part of a chain drive system.

You can calculate the speed of the driving pulley (driver) or driven pulley (follower) provided that both diameters and one speed are known. The same applies to gear and chain drives.

In the same way it is possible to work out the size of pulley or gear required for a power unit or machine to ensure that the equipment is driven at the correct operating speed. This assumes that the driving speed of the power unit is known; it can be checked in the instruction book or measured with a rev. counter. The operating speed of the driven machine must also be known.

To make any of these calculations, use this formula:

$$\frac{\text{Diameter of Follower}}{\text{Diameter of Driver}} = \frac{\text{Speed of Driver } (rpm)}{\text{Speed of Follower } (rpm)}$$

The formula can be moved around to find any unknown value provided the other three are known. For example, to find the speed of the follower:

$$\text{Speed of Follower } (rpm) = \frac{\text{Diameter of Driver} \times \text{Speed of Driver } (rpm)}{\text{Diameter of Follower}}$$

Example 1

Find the speed of a belt pulley which is 250 mm diameter and is driven by a 400 mm diameter pulley turning at 1,000 rpm.

$$\text{Speed of Follower} = \frac{400 \times 1{,}000}{250}$$

$$= 1{,}600 \text{ rpm.}$$

Example 2

Find the diameter of a driven pulley which will turn at 600 rpm if it is powered by a 9-inch pulley running at 420 rpm.

$$\text{Diameter of driven pulley} = \frac{\text{Diameter of Driver} \times \text{Speed of Driver } (rpm)}{\text{Speed of Follower } (rpm)}$$

$$= \frac{9 \times 420}{600}$$

$$= 6{\cdot}3 \text{ inches.}$$

In the same way, the formula can be used to calculate gear speeds and sizes.

Example 1
Find the speed of a gear wheel with 27 teeth when it is driven by a 36-tooth gear running at 90 rpm.

$$\text{Speed of Follower} = \frac{\text{Teeth on Driver} \times \text{Speed of Driver}}{\text{Teeth on Follower}}$$

$$= \frac{36 \times 90}{27}$$

$$= 120 \text{ rpm.}$$

Example 2
For a chain drive:
Find the speed of a shaft turned by a 48-tooth sprocket which is chain-driven by a 32-tooth sprocket running at 180 rpm.

$$\begin{array}{c}\text{Speed of Shaft}\\ \text{(Follower)}\end{array} = \frac{\text{Teeth on Driver} \times \text{Speed of Driver}}{\text{Teeth on Follower}}$$

$$= \frac{32 \times 180}{48}$$

$$= 120 \text{ rpm.}$$

Peripheral Speed

It is possible to change a pulley speed measured in rpm to peripheral or linear speed with another formula. Peripheral speed is the speed of a point on the circumference of a pulley which is more often used as a measurement for belt speed. A pulley speed will be measured in rpm but the speed of the belt it drives will be measured in feet per minute, metres per minute, etc. This means that the linear speed of a belt and the speed of a point on the rim of its driving pulley are the same.

An example of the use of linear speed is the speed of the rasp bars on the threshing cylinder of a combine harvester. The shaft speed of the cylinder may be 1,100 rpm to give a rasp bar speed of 6,000 feet per minute on a certain diameter cylinder. Another combine harvester with a different size cylinder will have a different speed in rpm to achieve the same threshing speed.

This formula is used to convert rpm to peripheral speed:

$$\underset{\substack{(\textit{feet per minute}\\ \text{or } \textit{metres per minute})}}{\text{Peripheral speed}} = \underset{(\textit{feet} \text{ or } \textit{metres})}{\text{Circumference}} \times \text{rpm}.$$

The circumference can be measured or can be found as follows:

$$\text{Circumference} = \pi \times \text{Diameter}$$

where

$$\pi = \frac{22}{7} \text{ or } 3{\cdot}142.$$

Example 1
Find the circumference of a wheel, 21 inches in diameter.

$$\text{Circumference} = \frac{22}{7} \times 21$$

$$= 66 \text{ inches.}$$

Now the peripheral speed of a 21-inch diameter pulley running at 60 rpm can be found:

$$\underset{(\textit{feet per min})}{\text{Peripheral speed}} = \text{circumference } (\textit{feet}) \times \text{rpm}$$

$$= \frac{66}{12} \times 60$$

$$= 330 \text{ feet per minute.}$$

(In this example the circumference of 66 inches is divided by 12 to convert this measurement to feet.)

Example 2
Find the peripheral speed of a 100 mm diameter pulley turning at 500 rpm in metres per minute.

$$\underset{(\textit{metres per minute})}{\text{Peripheral speed}} = \frac{22}{7} \times \frac{100}{1{,}000} \times 500$$

$$= 157 \text{ metres per minute.}$$

This can also be completed as a decimal calculation.

Example 3
Find the peripheral speed of a 400 mm pulley running at 500 rpm in metres per minute.

$$\text{Peripheral speed} = 3{\cdot}142 \times 0{\cdot}4\ (\text{metres}) \times 500$$
$$= 628 \text{ metres per minute.}$$

(Note that 400 mm has been changed to 0·4 m.)

ELECTRICITY

The unit of electrical power is the watt and this measurement is now also used as the unit of power for internal combustion engines. It is usual to refer to the power output in kilowatts using the symbol kw. There are 1,000 watts in a kilowatt.

In an electrical circuit it is possible to calculate the power in kw if the voltage of the circuit and the current flow in amps are known. In a similar way, the amperage or voltage of a circuit can be calculated.

A fuse is the safety valve in an electric circuit and it is important to use the correct size of fuse. You can calculate the current flow with this formula:

$$\text{Amps} = \frac{\text{Watts}}{\text{Volts}}$$

Power can be calculated by turning the formula round to give:

$$\text{Watts} = \text{Amps} \times \text{Volts.}$$

Example 1
Find the power in a 240 volt circuit with a current of 5 amps.

$$\text{Watts} = \text{Amps} \times \text{Volts}$$
$$= 5 \times 240$$
$$= 1{,}200 \text{ watts or } 1{\cdot}2 \text{ kw.}$$

Example 2
Find the current flow in a 240 volt circuit when running a 3 kw electric fire.

$$\text{Amps} = \frac{\text{Watts}}{\text{Volts}}$$

$$= \frac{3{,}000}{240} = 12{\cdot}5 \text{ amps.}$$

Fuses are rated in amps and are available in 3, 5, 10, and 13 amps when used in fused plugs. It is important to use the correct fuse in the plug for any item of electrical equipment.

Example:
Find the correct size of fuse for a 1 kw heater on a 240 volt circuit.

$$\text{Amps} = \frac{\text{Watts}}{\text{Volts}}$$

$$= \frac{1{,}000}{240} = 4{\cdot}17 \text{ amps.}$$

The correct size fuse will be 5 amps. Always use the nearest fuse rating *higher* than the calculated amperage.

Buying Electric Power

The consumer of electricity pays for the amount of electricity used plus a fixed charge. You can look at an electricity bill to find these costs. The fixed charge is set by the Electricity Board which also fixes the cost per unit.

One unit of electricity will operate a 1 kw appliance for one hour. With this information, you can calculate the running cost of any electric appliance if you know its power rating in watts.

Example:
Find the cost of running twelve electric light bulbs, each of 150 watts, for eight hours. The cost of 1 unit of electricity is 6·04 pence.

$$\text{Number of watts} = 12 \times 150 = 1{,}800 \text{ watts} = 1{\cdot}8 \text{ kw}$$

$$\text{Number of units in eight hours} = 1{\cdot}8 \text{ kw} \times 8 \text{ hours} = 14{\cdot}4 \text{ units}$$

$$\text{Cost} = 14{\cdot}4 \times 6{\cdot}04\text{p} = 86{\cdot}98 \text{ pence}$$

There is a relationship between horse power and watts. For the purpose of calculations:

$$1 \text{ horse power} = 746 \text{ watts.}$$

With this knowledge, fuse size and running costs for electric motors with power quoted in horse power can be calculated.

Example:
Find the correct size of fuse and the running cost for six hours for a 2 horse power electric motor. 1 unit of electricity costs 6·2 pence and the supply is 240 volts.

$$2 \text{ hp} = 2 \times 746 \text{ watts} = 1{,}492 \text{ watts}$$

$$\text{Amps} = \frac{\text{Watts}}{\text{Volts}}$$

$$= \frac{1{,}492}{240} = 6{\cdot}2 \text{ amps}$$

A 10 amp fuse will be required.

$$\text{Number of units} = 1{\cdot}492 \text{ kw} \times 6 \text{ hours}$$

$$= 8{\cdot}95 \text{ units}$$

$$\text{Cost} = 8{\cdot}95 \times 6{\cdot}2\text{p} = 55{\cdot}49 \text{ pence.}$$

ESTIMATING QUANTITIES

It is useful to be able to calculate the approximate quantities of material when carrying out such tasks as painting or building work. With a few simple facts, you can work out a variety of quantities.

Paint

Various types of paint will cover different areas according to their characteristics. For example:

1 litre of oil-based paint will cover about 17 square metres.

1 litre of non-drip jelly-type paint will cover about 12 square metres.

1 litre of emulsion paint will cover about 10 square metres.

So to find out how much paint is required, measure the area to be treated in square metres and divide this by the area covered by one litre of the material.

Example:
Find out how much emulsion paint is needed to paint a wall 12 metres long by 4 metres high.

$$\text{Wall area} = 12 \times 4 \text{ square metres}$$
$$= 48 \text{ square metres}$$

Emulsion paint has a coverage of 10 square metres per litre.

$$\text{Paint quantity} = \frac{48}{10} = 4{\cdot}8 \text{ litres}$$

5 litres will be required for a single coat.

Concrete

Home-mixed concrete needs deliveries of sand, stone or aggregate and cement. The quantities of each material will depend on the amount of concrete required and the proportions of each material in the mix. A typical mix for laying concrete paths is 1 part of cement, 2 parts of sand and 3 parts of aggregate. Quantities of sand and aggregate used are calculated and purchased in cubic metres. Some merchants sell these materials by weight.

To find the quantity of concrete required for a particular job, for example a concrete path, the total volume of the path must be calculated. This is done by using the dimensions for length, width and depth of concrete to be laid. Remember to use the same unit of measurement for all three dimensions. For a path 10 metres long, 1 metre wide and 100 millimetres deep, the depth dimension is first changed to 0·1 metres before working out the volume.

The quantities for 1 cubic metre mix of 1 part cement, 2 parts sand and 3 parts aggregate will be approximately as follows:

50 kg bags of cement – $6\frac{1}{4}$ bags
sand – 0·45 cubic metres
aggregate – 0·7 cubic metres

Example:
Find the quantity of concrete required to lay a path 16 m long, 1·2 m wide and 100 mm deep. What quantity of each constituent will be needed to mix this amount of concrete using 1 part cement, 2 parts sand and 3 parts aggregate?

Volume of concrete = $16 \times 1{\cdot}2 \times 0{\cdot}1$ cubic metres

= 1·92 cubic metres.

In practical terms, 2 cubic metres of concrete are required. Quantities of materials required are:

cement — $2 \times 6\frac{1}{4} = 12\frac{1}{2}$ bags

sand — $2 \times 0{\cdot}45\ m^3 = 0{\cdot}9\ m^3$

aggregate — $2 \times 0{\cdot}7\ m^3 = 1{\cdot}4\ m^3$.

Bricks and Blocks

When building a wall with bricks or blocks, it will of course be necessary to order the correct quantity of material. For a simple wall using bricks of 112 mm or $4\frac{1}{2}$ inch width, sixty bricks will be required for each square metre (or metre super) of wall. It is usual to add 10% to the order to allow for wastage and cutting. The nominal size (the effective dimensions including the mortar joint) of bricks is 225 mm × 75 mm × 112 mm.

Concrete blocks cover a greater area of wall space and have a nominal size of 450 mm × 225 mm. Nine blocks will provide 1 square metre of wall but here again, 10% should be added to the order.

Example 1
A brick wall 6 metres long, 4 metres high and 112 mm wide is required at a farm entrance. Calculate the number of bricks required for this wall allowing 25% extra for wastage and support piers.

Area of wall = 6 m × 4 m

= 24 square metres.

60 bricks are required per square metre.

(b) A 400 mm diameter pulley at 600 rpm in metres per minute.
(c) A 750 mm diameter pulley at 725 rpm in metres per minute.

16. Calculate the current flow and state the correct cartridge fuse size (3, 5, 10 or 13 amp), for the following items of electrical equipment when used on a 240 volt electricity supply.
(a) A 1,000 watt infra-red lamp.
(b) A two horse power electric motor.
(c) An inspection lamp with a 60 watt bulb.

17. Find the running cost for a 1 horse power electric motor and a two kilowatt heater when running continuously for 15 hours. The price of electricity is 6·2 pence per unit.

18. A new lighting system is to be installed in a farm building. There will be 6 lamp fittings with 150 watt bulbs, and a floodlight with a 500 watt bulb. Calculate the current flow in amps for this installation assuming that the power supply is 240 volts. What will be the hourly running cost of these lights if power costs 6·35 pence per unit?

19. Paint is to be ordered for a farm building. Emulsion paint will be used and two coats are needed. All of the interior walls of the building, 12 m long, 15 m wide and 3·5 m high, will be painted. There is a doorway, 4 m by 3 m, and 2 windows, each 2 m by 1·5 m. Calculate the minimum quantity of paint which must be ordered to decorate the walls (to the nearest litre).

20. A brick garage is to be built on a concrete base. It will have a base 150 mm deep, there will be no windows and one end will be fully taken up with the door. The walls will be 1 brick thick with supporting piers. The dimensions of the garage will be 5·2 m by 3·5 m wide, and the walls will be 4 m high. Calculate the quantity of material needed to mix the concrete using 1 part cement, 2 parts sand and 3 parts aggregate for the base and the number of bricks required for the walls. Allow 25% extra on the brick order for wastage and building the support piers.

Simple Conversions

YOUR calculator can provide the answer to all the calculations you normally have to make, but there are some occasions when mental arithmetic may be needed. If you were driving a car you would not have a spare hand to operate a calculator.

Britain measures road distances in miles; the rest of Europe in kilometres. It is very easy to convert in your head. One kilometre is $\frac{5}{8}$ of a mile, or 60% of a mile. Multiply the number of kilometres by 6 and discard the last figure to give an approximate conversion.

Thus 10 kilometres is 6 miles, 40 kilometres is 24 miles, 70 kilometres is 42 miles. This is a traveller's trick, and it is easy. Practise it.

Land measurement is in both acres and hectares. Some people have to use both if they are dealing with foreign visitors or travelling in other countries. One acre is 40% of a hectare. Multiply the number of acres by 4 and discard the last figure to give hectares.

Thus 10 acres is 4 hectares, 30 acres is 12 hectares, 150 acres is 60 hectares. These two examples are simple and quite accurate enough for most purposes. Other instant calculations of this type can be done, but may not be accurate enough. You can try some of them.

One pound is roughly half a kilo, or more accurately 45%. You can multiply by 9, then divide by 2 and discard the last figure to give the weight in kilos. Test this out.

In the same way, one litre is roughly 20% of a gallon. To convert to gallons you can multiply litres by 2 and discard the last figure. Test this out.

One foot is roughly $\frac{1}{3}$ of a metre, or more accurately 30%. Multiply feet by 3 and discard the last figure to give the number of metres. Test this out.

Ten centimetres is almost exactly 4 inches (the old measurement of a *hand*, used to measure the height of horses). This is a simple and useful conversion, for dealing with measurements such as 20, 30 or 40 centimetres or those expressed as parts of a metre, such as 0·5, 0·6, 0·8 m. Try these out.

The authors agree that for accuracy it is better to put a decimal point before the last figure (after working out the sum) and round it up or down.

Example: 28 kilometres × 6 = 16·8 miles.

This shows you how to do it. Now you can design your own simple conversions.

To make accurate conversions, use your calculator and refer to the tables of conversion which follow this section. Occasionally check and double-check, by reversing the calculation. It is easy to misplace a decimal point.

Body Measurements

We have already referred to the *hand* used for centuries in Britain to measure horses. In many parts of the world, the hand, the foot, the forearm, and the thumb (one inch) are used – even to the extent of the thumb being used in Scandinavia to measure timber. There is an old method for measuring cloth from the end of the outstretched hand to your own nose. Try it.

While we do not suggest such measurements for accuracy, it is not a bad thing to check some of your own personal measurements, in case you need them for rough and ready measurements of length at a time when you have no measuring tape or rule. Try this out with such body measurements as: foot, forearm (the cubit of the Bible), pace, double arm's length and finger span (thumb to little finger).

You will find it interesting – perhaps it may be useful – to take your own body measurements and put the results in the small table below. It may even have some historical value one day.

	Metric	*British*
Foot		
Pace		
Forearm		
Double arm's length		
Finger span		
Round your head		

Basic Tables and Conversions

LENGTH

METRIC

kilometre (km)	= 1,000 metres
metre (m)	
centimetre (cm)	= 0·01 metre
millimetre (mm)	= 0·001 metre
1 millimetre	= 0·0394 in
1 centimetre	= 0·394 in
1 metre	= 1·09 yd
1 kilometre	= 0·621 miles

BRITISH

1 inch (in)	
1 foot (ft)	= 12 inches
1 yard (yd)	= 3 feet
1 mile	= 1,760 yards
1 inch	= 2·54 cm or 25·4 mm
1 foot	= 0·30 m
1 yard	= 0·91 m
1 mile	= 1·61 km

CONVERSION FACTORS

centimetres to in	× 0·394
millimetres to in	× 0·0394
metres to ft	× 3·29
metres to yd	× 1·09
kilometres to miles	× 0·621
inches to cm	× 2·54
or mm	× 25·4
feet to m	× 0·305
yards to m	× 0·914
miles to km	× 1·61

AREA

METRIC

hectare (ha)	= 10,000 square metres
square metre (m^2)	
square centimetre (cm^2)	= 0·0001 square metre
1 sq centimetre	= 0·16 sq in
1 sq metre	= 1·20 sq yd
1 sq metre	= 10·8 sq ft
1 hectare	= 2·47 ac

BRITISH

1 sq inch (sq in)	
1 sq foot (sq ft)	= 144 sq inches
1 sq yard (sq yd)	= 9 sq feet
1 acre (ac)	= 4,840 sq yards
1 sq inch	= 6·45 cm^2
1 sq foot	= 0·093 m^2
1 sq yard	= 0·836 m^2
1 acre	= 4,047 m^2 or 0·405 ha

CONVERSION FACTORS

sq metres to sq ft	× 10·8
sq metres to sq yds	× 1·20
hectares to acres	× 2·47
sq feet to m^2	× 0·0929
sq yards to m^2	× 0·836
acres to hectares	× 0·405

VOLUME and CAPACITY

METRIC

cubic metre (m^3)	= 1 kilolitre = 1,000 litres
litre (l)	
millilitre (ml)	= 1 cubic centimetre (cc) or 0·001 litre
100 millilitres	= 0·176 pint
1 litre	= 1·76 pints
1 kilolitre	= 220 gallons

BRITISH

1 fluid once (fl oz)	
1 pint	= 20 fl oz
1 gallon	= 8 pints
1 fluid once	= 28·4 ml
1 pint	= 0·568 litres
1 gallon	= 4·55 litres

CONVERSION FACTORS

litres to pints	× 1·76
litres to gallons	× 0·220
pints to litres	× 0·568
gallons to litres	× 4·55

TEMPERATURE

(°C) degree Centigrade (also called Celsius)
Freezing point = 0°C
Boiling point = 100°C

(°F) degree Fahrenheit
Freezing point = 32°F
Boiling point = 212°F

Conversion

(°C × 1·8) + 32 = °F

(°F − 32) ÷ 1·8 = °C

WEIGHT

METRIC

metric tonne (tonne)	= 1,000 kilograms
kilogram (kg)	
gram (g)	= 0·001 kilogram
milligram (mg)	= 0·001 gram
1 gram	= 0·035 oz
100 grams	= 3·53 oz
1 kilogram	= 2·20 lb
1 tonne	= 2,204 lb or 0·984 ton

BRITISH

1 ounce (oz)	
1 pound (lb)	= 16 ounces
1 hundredweight (cwt)	= 112 pounds
1 ton	= 20 hundredweights
1 ounce	= 28·3 g
1 pound	= 454 g or 0·454 kg
1 hundredweight	= 50·8 kg
1 ton	= 1,016 kg or 1·016 tonne

CONVERSION FACTORS

grams to oz	× 0·0353
grams to lb	× 0·00220
kilograms to lb	× 2·20
kilograms to cwt	× 0·0197
tonnes to tons	× 0·984

ounces to g	× 28·3
pounds to g	× 454
pounds to kg	× 0·454
hundredweights to kg	× 50·8
hundredweights to tonnes	× 0·0508
tons to kg	× 1,016
tons to tonnes	× 1·016

Answers

General Problems

1. 26·39; 16·46; 51·08.
2. 7·52; 0·01 (0·009); 25·52.
3. 10·22 cm; 6·36 cm
   4·0 m; 40·38 m^2
   12·0 mm; 193·30 mm^3
   0·75 m; 0·60 m
   5·68 cm; 159·14 cm^3.
4. 0·63; 0·35; 3·44; 70·30; 0·06.
5. 1·59 cm.
6. 13,386·92.
7. 42·72.
8. 0·712 mm.
9. 2,152 rpm.
10. £6,094; no, the % increase was 17·78%.

Farm Crop Problems

1. 2·8 m.
2. 182 m.
3. 387 rows.
4. 19 tile drains.
5. 23·75 m^2; 17·28 m.
6. 5·4 m^3; 5,400 litres.
7. 574 litres; 862 kg.
8. 42 tonnes; 38 tonnes; 28 tonnes.
9. Hay 167 kg, 109 kg; wheat straw 103 kg, 57 kg; barley straw 87 kg, 44 kg; oat straw 95 kg, 52 kg; silage 667 kg, all per cubic metre (m^3).
10. 558 tonnes.
11. 596 tonnes; 21% silage is 125 tonnes DM; 27% silage is 146 tonnes DM.
12. No, the store could only hold 116 tonnes.
13. 25 tonnes.
14. 655·6 cu metres; 3 for wheat; 3 for barley; 1 for oats; Total 7.
15. 10·12 tonnes; 1·1 tonnes less.

16. 260 plants survived over winter.
17. 2·2 m row length; 53 tonnes per ha.
18. Wheat: compound A, 4·2 × 50 kg per ha plus 34·5% N at 2·9 × 50 kg and 7·25 × 50 kg per ha. Barley: compound C, 8·2 × 50 kg per ha. Sugar beet: compound G, 15·3 × 50 kg per ha. Potatoes: compound F, 25·3 × 50 kg per ha. Grass: 34·5% N at 4·9 × 50 kg and compound F, 6·7 × 50 kg per ha.
19. (a) 38·89p; (b) 28·89p.
20. 495 tonnes.
21. 293 kg of N and 140 kg of P per ha.
22. 102 kg N; 51 kg P; 69 kg K; 54 journeys.
23. 8 tonnes; 12 tonnes.
24. Product A (A costs £11.25, B costs £12.00).
25. £129 per can.
26. 11 cans; 12 cans.
27. £3,150·21.
28. (i) 18%; (ii) 9%.
29. (a) 25; (b) 140.
30. (a) 3,175,000 litres; (b) 13,716,000 litres.

Farm Livestock Problems

1. 272·6 livestock units.
2. 44 bullocks.
3. 48 ha.
4. 29 metres; 6·6 days.
5. 23·63 tonnes; 104·63 tonnes; 675 tonnes.
6. 185 days.
7. 150 mm.
8. 563 kg; 113 kg.
9. 27·6 tonnes.
10. 0·7 kg.
11. 10·5 kg; 14 kg; 17·5 kg; 2·1 kg.
12. 27·3 tonnes; 142 tonnes.
13. 419·5 tonnes.
14. 825 kg.
15. 180 tonnes; 153 tonnes of barley.
16. 414,880 litres.
17. 588 lambs.
18. The better result is (a). [(a) yields 159·7%; (b) yields 149·4%].
19. 21·6 piglets per annum.
20. 0·52 kg per day; 3·17 kg of food.
21. 1·06 kg per day.
22. 63 kg; 297 kg; 33 kg.

23. 75·7%; 69·9%; heavier pig is better.
24. 4·07 kg per kg liveweight gain; 0·95 tonnes barley.
25. 2·5 kg; 3·8 kg; 3·16 kg of food per kg L.W.G.
26. 49·75p; 66·32p; 58·26p.
27. £617·34.
28. £11,499·95.
29. Liveweight £60.48; Deadweight £55.22.
30. 53·27 Grade 2 pigs.

HORTICULTURAL PROBLEMS

1. 10·46 m^2.
2. 120 m × 180 m.
3. 6 m^2; 4 times.
4. (a) 200; (b) 222; 800.
5. 600; £396.
6. 12·72 m^2.
7. 5 m by 5 m.
8. 4·5 m square.
9. 3 rolls; £126 but not all is used.
10. 92,736 watts (or 92·74 kilowatts).
11. 27·8°C.
12. 45°F.
13. 2,530 m^3; 9 fumigants.
14. Compound F at 24·7 × 50 kg; Compound A at 4·2 × 50 kg; Compound G at 18·8 × 50 kg; Compound B at 14 × 50 kg plus 8·1 × 50 kg of 34·5% N.
15. 2·23 tonnes ground limestone; 1·11 tonnes burnt lime; 1·49 tonnes slaked lime.
16. 33 plants.
17. 59 tubers.
18. 121·6 m^2.
19. 210·8 m^2.
20. 810 m^2.
21. Poa 20%, Festuca 70%, Agrostis 10%.
22. 210 ml.
23. (a) 181·8 g; (b) 349 ppm of P_2O_5.

MACHINERY PROBLEMS

1. 24·2 turns.
2. 25·25 turns.
3. 6 minutes.
4. 1·65 minutes.

5. 5·2 minutes.
6. 2·3 minutes.
7. 880 yards.
8. 476 metres.
9. (a) 6 acres per hour.
   (b) 34·7 acres per hour.
   (c) 18 acres per hour.
10. (a) 17·8%.
    (b) 18·7%.
    (c) 15·9%.
11. 4·2 inches.
12. 17·1 inches.
13. 365 rpm.
14. 74·6 rpm.
15. (a) 1,260 feet per minute.
    (b) 754 metres per minute.
    (c) 1,710 metres per minute.
16. (a) 4·2 amps, 5 amp fuse.
    (b) 6·2 amps, 10 amp fuse.
    (c) 0·25 amps, 3 amp fuse.
17. £2·55.
18. 5·8 amps; 8·89p per hour.
19. 34 litres.
20. 17 bags of cement, 1·2 cubic metres of sand; 1·9 cubic metres of aggregate; 4,170 bricks.

Index

FARMING PRESS BOOKS

The following are samples from the wide range of agricultural and veterinary books published by Farming Press. For more information or for a free illustrated book list please contact:

Farming Press Books, 4 Friars Courtyard,
30–32 Princes Street, Ipswich IP1 1RJ, United Kingdom
Telephone (0473) 43011

Farm Building Construction MAURICE BARNES AND CLIVE MANDER
Covers all aspects of farm building work with details on blockwork, brickwork, timber, flooring, walls etc., for new constructions or improvements.

Farm Machinery – Third Edition BRIAN BELL
Completely revised to include photographs, with new material on harvesting machinery, all-terrain vehicles, rough-terrain forklift trucks and estate maintenance equipment.

Farm Livestock GRAHAM BOATFIELD
An excellent introduction to livestock production methods designed for those taking their first steps in agriculture. Full details on cattle, sheep and pigs, with general sections on breeding, feeding and health.

Farm Crops GRAHAM BOATFIELD
Provides an outline for farm crops and crop husbandry along with basic knowledge and scientific principles underlying the growing of these crops.

The Modern Shepherd SAM MEADOWCROFT AND DAVE BROWN
The authors show how a farmer can double lamb output by combining first-rate shepherding with the best of the techniques developed over the last two decades.

Cereal Husbandry E. JOHN WIBBERLEY
A panoramic view of temperate cereal production focusing in particular on wheat, barley, oats and rye with emphasis on the practical application of principles. An invaluable reference book for farmers and students

Farming Press also publish four monthly magazines:
Livestock Farming, *Dairy Farmer*, *Pig Farming* and *Arable Farming*. For a specimen copy of any of these magazines please contact Farming Press at the address above.